普通高等教育软件工程专业"十二五"规划教材

# JSP 程序设计

主编 于 彬

副主编 陈 玮 张志峰 张 丽

科学出版社

北 京

## 内 容 简 介

本书遵循以语言为基础、以应用为主导的编写原则，理论联系实际，并通过大量的实例，循序渐进地介绍了有关 JSP 开发所涉及的各类知识。

全书共 11 章，首先从最基础的 JSP 程序设计基础知识和 JSP 开发环境的搭建开始，然后介绍 JSP 基础、内置对象、和 Servlet 结合、和 JavaBean 结合、JSTL 和 AJAX 技术等，最后通过案例开发讲解 JSP 知识的综合应用。本书对基础知识介绍详细，理论联系实际，具有很强的操作性。书中还提供了大量通过测试的、可运行的完整实例代码，这些实例都有相应的设计步骤、代码详解、程序运行结果等，使读者能够快速掌握和运用 JSP 编程技巧。

本书可作为大学计算机软件及相关专业的 Web 编程课程教材，也可作为技术支持人员和程序开发人员的参考书。

---

图书在版编目（CIP）数据

JSP 程序设计 / 于彬主编．—北京：科学出版社，2013
普通高等教育软件工程专业"十二五"规划教材
ISBN 978-7-03-037564-3

Ⅰ．①J… Ⅱ．①于… Ⅲ．①Java 语言—网页制作工具 Ⅳ．①TP312 ②TP393.092

中国版本图书馆 CIP 数据核字（2013）第 110103 号

责任编辑：于海云 / 责任校对：李 影
责任印制：闫 磊 / 封面设计：迷底书装

**科 学 出 版 社** 出版
北京东黄城根北街 16 号
邮政编码：100717
http://www.sciencep.com

保定市中画美凯印刷有限公司 印刷
科学出版社发行 各地新华书店经销

\*

2013 年 6 月第 一 版　开本：787×1092 1/16
2013 年 6 月第一次印刷　印张：20
　　　　　　　　　　　字数：506 000
**定价：45.00 元**
（如有印装质量问题，我社负责调换）

# 普通高等教育软件工程专业"十二五"规划教材

## 编 委 会

**主任委员**
　　李占波　　郑州大学软件技术学院副院长

**副主任委员**
　　车战斌　　中原工学院软件学院院长
　　刘黎明　　南阳理工学院软件学院院长
　　刘建华　　华北水利水电大学软件学院院长
　　乔保军　　河南大学软件学院副院长

**委　员**
　　高　岩　　河南理工大学计算机科学与技术学院副院长
　　邓璐娟　　郑州轻工业学院软件学院副院长
　　史玉珍　　平顶山学院软件学院副院长
　　周文刚　　周口师范学院计算机科学与技术学院副院长
　　席　磊　　河南农业大学信息与管理科学学院系主任
　　陈建辉　　郑州航空工业管理学院计算机科学与应用系副主任
　　张永强　　河南财经政法大学计算机与信息工程学院副院长
　　郑延斌　　河南师范大学计算机与信息工程学院副院长
　　谭营军　　河南职业技术学院信息工程系副主任
　　赵素萍　　洛阳师范学院信息技术学院软件工程系主任
　　潘　红　　新乡学院计算机与信息工程学院院长

# 《JSP 程序设计》编委会

主　编　于　彬

副主编　陈　玮　张志峰　张　丽

参　编　王　倩　熊蜀峰　宋振方　屠晓云

# 前　言

本书是为了区别于传统的教材而编写的适应软件工程专业高等教育、培养应用型本科专业人才需要的教材。在编写方案制定上力求体现教育部的教育教学改革精神，树立以素质教育为基础、能力为本位、就业为导向的"以项目为驱动的教学模式"改革新观念。教材编写紧紧围绕高等院校的软件工程专业培养目标，突出应用型本科人才培养模式；在内容体系上打破科学性，突出实用性和技能性；理论部分坚持以够用为度，以应用为主旨，强调实践、实训教学环节，培养学生的综合实践能力。

本书的读者应具有一定的 Java 语言基础。通过本书的学习，学生能够掌握 Java Web 开发的基本知识和主要技术。本书各章节的例子都是从简单到复杂，逐步深入。叙述上追求简单直观，通俗易懂，便于读者学习掌握 JSP 技术。

全书共 11 章。第 1 章介绍了 JSP 程序设计基础知识；第 2 章介绍了 Java Web 环境配置等内容；第 3 章主要介绍 JSP 基础知识；第 4 章详细地介绍了 JSP 的内置对象；第 5 章介绍了 Servlet 技术，该技术是非常重要的一种组件技术，能极大地发挥 Java 语言的优点，Servlet 被称为 Java 组件技术的核心；第 6 章介绍了 JavaBean 组件技术以及 JavaBean 在 Java Web 开发中的应用；第 7 章介绍了在 JSP 页面中如何实现对文件的操作；第 8 章介绍了 JDBC 技术；第 9 章介绍了 JSTL 技术；第 10 章介绍了 AJAX 异步交互技术；第 11 章通过两个比较典型的实例，综合运用相关 Java Web 技术，介绍了如何使用 JSP+JavaBean 模式和 JSP+JavaBean+Servlet 模式来开发具体的 Web 应用系统。

本书配备有教学课件、教学大纲、教学日历、代码、模拟考试试题，以及课后习题参考答案。

本书由于彬担任主编。其中第 1、10 章由河南农业大学张丽编写，第 2、7 章由南阳理工学院软件学院于彬编写，第 3、9 章由河南理工大学计算机学院陈玮编写，第 4 章由河南大学软件学院王倩编写，第 5 章由平顶山学院软件学院熊蜀峰编写，第 6 章由洛阳师范学院信息技术学院宋振方编写，第 8 章由郑州轻工业学院软件学院张志锋编写，第 11 章由南阳理工学院软件学院屠晓云编写。

最后感谢科学出版社、郑州轻工业学院、河南理工大学、河南农业大学、河南大学、平顶山学院、洛阳师范学院、南阳理工学院给予的大力支持。

由于时间仓促，加之编者水平有限，书中难免存在不足之处，恳请读者批评指正。

编　者

2013 年 5 月

# 目 录

前言

## 第1章 JSP 程序设计基础 ............ 1
### 1.1 Web 应用程序 ............ 1
#### 1.1.1 Web 技术的由来与发展 ............ 1
#### 1.1.2 Web 动态网页技术 ............ 3
#### 1.1.3 Web 应用程序的工作原理 ............ 4
### 1.2 JSP 技术概述 ............ 5
### 1.3 JSP 的工作原理 ............ 6
### 1.4 JSP 程序开发模式 ............ 7
#### 1.4.1 JSP 两种体系结构 ............ 7
#### 1.4.2 JSP 开发 Java Web 站点的主要方式 ............ 8
### 1.5 小结 ............ 8
### 1.6 习题 ............ 9

## 第2章 JSP 开发环境搭建 ............ 10
### 2.1 JSP 的运行环境 ............ 10
#### 2.1.1 JSP 的运行环境 ............ 10
#### 2.1.2 JDK 的安装与配置 ............ 11
#### 2.1.3 Tomcat 的安装与启动 ............ 15
#### 2.1.4 JSP 开发工具——Eclipse ............ 19
### 2.2 Java Web 项目的建立与部署 ............ 22
#### 2.2.1 Java Web 项目的建立 ............ 22
#### 2.2.2 创建一个 JSP ............ 25
#### 2.2.3 创建 Servlet ............ 27
#### 2.2.4 Java Web 项目的部署 ............ 31
### 2.3 BBS 案例的分析与设计 ............ 32
#### 2.3.1 系统分析 ............ 32
#### 2.3.2 系统设计 ............ 33
### 2.4 小结 ............ 36

## 第3章 JSP 基础 ............ 37
### 3.1 JSP 简介 ............ 37
#### 3.1.1 JSP 页面的基本结构 ............ 37
#### 3.1.2 JSP 的运行原理 ............ 38
### 3.2 JSP 语法 ............ 42

3.2.1 变量和方法的声明 ................................................ 42
  3.2.2 JSP 表达式 ........................................................ 43
  3.2.3 Java 程序片（scriptlet）........................................ 44
  3.2.4 JSP 注释 ............................................................ 45
 3.3 JSP 指令标签 ............................................................ 47
  3.3.1 page 指令 .......................................................... 47
  3.3.2 include 指令 ....................................................... 49
 3.4 JSP 动作标签 ............................................................ 51
  3.4.1 jsp:include 动作标签 ............................................ 51
  3.4.2 jsp:forward 动作标签 ........................................... 51
  3.4.3 param 动作标签 .................................................. 52
  3.4.4 useBean 动作标签 ............................................... 54
 3.5 小结 ........................................................................ 56
 3.6 习题 ........................................................................ 56
第 4 章 JSP 内建对象 ............................................................. 57
 4.1 out 对象 ................................................................... 57
 4.2 request 对象 ............................................................. 60
  4.2.1 获取 HTML 表单提交的数据 ................................... 61
  4.2.2 汉字信息处理 ..................................................... 66
  4.2.3 常用方法举例 ..................................................... 67
 4.3 response 对象 ........................................................... 70
  4.3.1 动态响应 contentType 属性 ................................... 70
  4.3.2 response 实现网页的自动刷新 ................................ 71
  4.3.3 response 重定向 .................................................. 72
  4.3.4 response 的状态行 ............................................... 72
 4.4 session 对象 ............................................................. 73
  4.4.1 session 对象的 ID ................................................ 74
  4.4.2 session 对象与 URL 重写 ...................................... 75
  4.4.3 session 对象的常用方法 ........................................ 77
 4.5 application 对象 ........................................................ 80
  4.5.1 application 对象的常用方法 ................................... 80
  4.5.2 用 application 对象制作留言板 ................................ 81
 4.6 小结 ........................................................................ 84
 4.7 习题 ........................................................................ 84
第 5 章 Servlet .................................................................... 85
 5.1 Servlet 概述 .............................................................. 85
  5.1.1 Servlet 简介和优点 .............................................. 85
  5.1.2 Servlet 与 JSP 的关系 ........................................... 86
  5.1.3 JSP 文件编译过程 ................................................ 86

  5.1.4 HTTP 基础知识 90
 5.2 Servlet 的编译和运行 92
  5.2.1 一个简单的 Servlet 例子 92
  5.2.2 存放 Servlet 的目录 93
  5.2.3 运行 Servlet 94
 5.3 Servlet 的体系结构 95
  5.3.1 一个基本 Servlet 程序的组成 95
  5.3.2 Servlet 应用程序体系结构 96
  5.3.3 Servlet 层次结构 96
 5.4 Servlet 的生命周期 97
  5.4.1 Servlet 的生命周期 97
  5.4.2 Servlet 的基本方法 99
 5.5 JSP 和 Servlet 的交互 101
  5.5.1 通过表单向 Servlet 提交数据 101
  5.5.2 从 Servlet 到 JSP 的信息传递 106
 5.6 Servlet 的高级应用 109
  5.6.1 Servlet 的初始化参数 109
  5.6.2 过滤器 112
  5.6.3 监听器 117
 5.7 小结 122
 5.8 习题 122

# 第 6 章 JavaBean 组件技术 124
 6.1 JavaBean 概述 124
 6.2 JavaBean 的编写和使用 124
  6.2.1 编写 JavaBean 124
  6.2.2 使用 JavaBean 126
  6.2.3 设置 JavaBean 属性 127
  6.2.4 获得 JavaBean 属性 134
  6.2.5 设置 JavaBean 的范围 136
 6.3 Java Web 开发模型 147
  6.3.1 JSP 和 JavaBean 开发模型 147
  6.3.2 JSP+JavaBean+Servlet 开发模型 151
  6.3.3 两种开发模型比较 156
 6.4 分页 JavaBean 开发 156
 6.5 小结 166
 6.6 习题 166

# 第 7 章 JSP 中的文件操作 167
 7.1 File 类和数据流 167
  7.1.1 File 类 167

7.1.2 字节输入流类和字节输出流类 ································· 171
7.2 文件上传 ································································· 173
　　7.2.1 JSP 页面处理文件上传 ········································· 173
　　7.2.2 Servlet 处理文件上传 ·········································· 175
7.3 文件下载 ································································· 178
7.4 小结 ······································································ 180
7.5 习题 ······································································ 180

## 第 8 章 JDBC 与数据库访问 ················································ 181
8.1 JDBC 基础知识 ························································· 181
　　8.1.1 JDBC 简介 ······················································ 181
　　8.1.2 DriverManager ················································· 184
　　8.1.3 Connection ······················································ 185
　　8.1.4 Statement ························································ 186
　　8.1.5 ResultSet ························································ 187
8.2 使用 JDBC 访问数据库 ·············································· 187
　　8.2.1 使用 JDBC 访问数据库的一般步骤 ························ 187
　　8.2.2 使用 JDBC 驱动访问 MySQL 数据库 ····················· 189
　　8.2.3 访问 Microsoft SQL Server 2000 数据库及其应用实例 ···· 193
　　8.2.4 访问 Microsoft SQL Server 2008 数据库及其应用实例 ···· 197
8.3 数据库的增、删、改、查操作 ······································· 203
　　8.3.1 数据库的增、删、改、查操作 ······························ 203
　　8.3.2 基于 MVC 模式的学生信息管理系统 ····················· 205
8.4 JSP 在数据库应用中的其他相关问题 ····························· 227
　　8.4.1 分页技术 ························································ 227
　　8.4.2 常见中文乱码处理方式 ······································ 229
8.5 小结 ······································································ 231
8.6 习题 ······································································ 231

## 第 9 章 JSP 标准标签库 ······················································ 232
9.1 JSP 标准标签库简介 ·················································· 232
　　9.1.1 概述 ······························································ 232
　　9.1.2 JSTL 的使用 ···················································· 232
9.2 核心标签库 ······························································ 233
9.3 国际化标签库 ··························································· 237
9.4 数据库标签库 ··························································· 241
9.5 XML 标签库 ···························································· 244
9.6 函数标签库 ······························································ 248
9.7 小结 ······································································ 251
9.8 习题 ······································································ 251

# 第 10 章 AJAX ... 252
## 10.1 AJAX 简介 ... 252
### 10.1.1 AJAX 包含的技术 ... 252
### 10.1.2 AJAX 的运行原理 ... 253
## 10.2 AJAX 开发框架 ... 256
## 10.3 AJAX 应用 ... 259
### 10.3.1 基于 JSP 的 AJAX ... 259
### 10.3.2 基于 Servlet 的 AJAX ... 265
### 10.3.3 AJAX 的应用场景 ... 273
## 10.4 小结 ... 274
## 10.5 习题 ... 274

# 第 11 章 综合案例 ... 275
## 11.1 博客网站 ... 275
### 11.1.1 系统功能 ... 275
### 11.1.2 数据库设计 ... 275
### 11.1.3 系统主要功能的实现 ... 276
## 11.2 网上书店 ... 278
### 11.2.1 系统功能 ... 278
### 11.2.2 数据库设计 ... 278
### 11.2.3 系统的关键技术 ... 279
### 11.2.4 各个页面的设计 ... 281
## 11.3 小结 ... 307
## 11.4 习题 ... 307

# 参考文献 ... 308

# 第 1 章　JSP 程序设计基础

当今对 Web 技术的强大需求推进着各种 Web 技术的应运而生。本章主要讲解 Web 技术的相关概念与原理，主要内容包括：Web 应用程序、JSP 技术概述、JSP 工作原理、JSP 程序开发模式。

## 1.1　Web 应用程序

随着信息化时代的到来，我们对网络的依赖越来越多，我们从网络上获取许多的信息资源。作为信息传送的主题，Web 受到越来越多人的青睐。

### 1.1.1　Web 技术的由来与发展

Web（World Wide Web，简称 WWW 或者 Web）是由蒂姆·伯纳斯-李（Tim Berners-Lee，万维网之父，1955 年出生于英国，不列颠帝国勋章获得者，英国皇家学会会员，英国皇家工程师学会会员，美国国家科学院院士）于 1989 年 3 月提出的万维网设想而发展起来的。1990 年 12 月 25 日，他在日内瓦的欧洲粒子物理实验室里开发出了世界上第一个网页浏览器。他是关注万维网发展的万维网联盟的创始人，并获得世界多国授予的各种荣誉。他最杰出的成就是免费把万维网的构想推广到全世界，让万维网科技获得迅速的发展，并极大地改变了人类的生活方式。

国际互联网 Internet 在 20 世纪 60 年代就诞生了，为什么没有迅速流传开来呢？其实，很重要的原因是因为连接到 Internet 需要经过一系列复杂的操作，网络的权限也很分明，而且网上内容的表现形式极其单调枯燥。Web 通过一种超文本方式把网络上不同计算机内的信息有机地结合在一起，并且可以通过超文本传输协议（HTTP）从一台 Web 服务器转到另一台 Web 服务器上检索信息。Web 服务器能发布图文并茂的信息，在软件支持下还可以发布音频和视频信息。此外，Internet 的许多其他功能，如 E-mail、Telnet、FTP 等都可通过 Web 实现。美国著名的信息专家尼葛洛庞帝教授认为：1989 年是 Internet 历史上划时代的分水岭。Web 技术确实给 Internet 赋予了强大的生命力，Web 浏览的方式给互联网打开了新局面。

Web 的前身是 1980 年由蒂姆·伯纳斯-李负责的一个项目演变而来的。1990 年第一个 Web 服务器开始运行。1991 年，CERN（Conseil Européen pour la Recherche Nucléaire，欧洲核子研究组织）正式发布了 Web 技术标准。W3C（World Wide Web Consortium，万维网联盟或者 W3C 理事会）于 1994 年 10 月由蒂姆·伯纳斯-李在麻省理工学院计算机科学实验室成立，负责组织、管理和维护 Web 相关的各种技术标准，目前的 Web 版本是 Web 3.0。

早期的 Web 应用主要是使用 HTML 语言编写、运行在服务器端的静态页面。用户通过浏览器向服务器端的 Web 页面发出请求，服务器端的 Web 应用程序接收到用户发送的请求后，读取地址所标识的资源，加上消息报头把用户访问的 HTML 页面发送给客户端的浏览器。

HTML（Hypertext Markup Language，超文本标记语言）是一种描述文档结构的语言，不能描述实际的表现形式。HTML 的历史最早可以追溯到 1945 年。1945 年，范内瓦·布什

(Vannevar Bush)提出了文本和文本之间通过超级链接相互关联的思想,并给出设计方案。范内瓦·布什是具有6个不同学位的科学家、教育家和政府官员,与20世纪许多著名的事件都有关系,如组织和领导了制造第一颗原子弹的著名的"曼哈顿计划"、氢弹的发明、登月飞行、"星球大战计划"等。正如历史学家迈克尔·雪利所言:"要理解比尔·盖茨和比尔·克林顿的世界,你必须首先认识范内瓦·布什。"正是因其在信息技术领域多方面的贡献和超人远见,范内瓦·布什获得了"信息时代的教父"的美誉。1960年这种信息关联技术被正式命名为超文本技术。从1991年HTML语言正式诞生以来推出了多个不同的版本,其中对Web技术发展具有重大影响的主要有两个版本:1996年推出的HTML 3.2和1998年推出的HTML 4.0。1999年W3C颁布了HTML 4.0.1。目前大多数Web服务器和浏览器等相关软件均支持HTML 4.0.1标准。HTML V5版本将拥有更大的应用空间。

但是让HTML页面丰富多彩、动感无限的是CSS(Cascading Style Sheets,级联样式表)和DHTML(Dynamic HTML,动态HTML)技术。1996年底,W3C提出了CSS标准,CSS大大提高了开发者对信息展现格式的控制能力。DHTML技术无需启动Java虚拟机或其他脚本环境,在浏览器的支持下,获得更好的展现效果和更高的执行效率。

最初的HTML语言只能在浏览器中展现静态的文本或图像信息,这远不能满足人们对信息丰富性和多样性的强烈需求。这就促使Web技术由静态技术向动态技术转化。

第一种真正使服务器能根据运行时的具体情况,动态生成HTML页面的技术是CGI(Common Gateway Interface,公共网关接口)技术。1993年,CGI 1.0的标准草案由NCSA(National Center for Supercomputing Applications,国家计算机安全中心)提出。1995年,NCSA开始制定CGI 1.1标准。CGI技术允许服务端的应用程序根据客户端的请求,动态生成HTML页面,这使客户端和服务端的动态信息交换成为了可能。随着CGI技术的普及,聊天室、论坛、电子商务、信息查询、全文检索等各式各样的Web应用蓬勃兴起,人们终于可以享受到信息检索、信息交换、信息处理等更为便捷的信息服务了。

CGI是Web服务器扩展机制,它允许用户调用Web服务器上的CGI程序。用户通过单击某个链接或者直接在浏览器的地址栏中输入URL访问CGI程序,Web服务器接收到请求后,发现该请求是给某个CGI程序,就启动并运行该CGI程序,对用户请求进行处理。CGI程序解析请求中的CGI数据,处理数据,并产生一个响应(HTML页面)。该响应被返回Web服务器,Web服务器包装该响应,如添加报头消息,以HTTP响应的形式发送给客户端浏览器。

但是,用户在使用CGI程序时候发现编写程序比较困难,而且对用户请求和响应时间较长。由于CGI程序的这些缺点,开发人员需要其他的CGI方案。

1994年,Rasmus Lerdorf发明了专用于Web服务端编程的PHP(Personal Home Page,个人网页)语言。它开始是一个用Perl语言编写的简单的程序,Rasmus Lerdorf用它来与访问其主页的人保持联系。后来它逐步流行,Rasmus又重新写了整个解析器,并命名为PHP V1.0,当然功能还不是十分完善。此后其他程序员开始参与PHP源码的编写,1997年,Zeev Suraski和Andi Gutamns又重新编写了解析器,使功能基本完善,随后发布了PHP 3。

PHP程序可以运行在UNIX、Linux或者Windows操作系统下,对客户端浏览器也没有特殊的要求。PHP也是将脚本描述语言嵌入HTML文档中,它大量采用了C、Java和Perl语言的语法,并加入了各种PHP自己的特征。PHP的优点如下:首先它是免费的,这对于许多要考虑运行成本的商业网站来说尤其重要;其次它开放源代码,因为这一点,所以才会有很多爱好者不断发展它，使之更具有生命力；再次它有多平台支持，可以运行在所有的操作系统之下；最后它效率高，同 ASP 相比，PHP 占用资源少，执行速度较快。

1996 年，微软公司（微软公司是世界个人计算机软件开发的先导，由比尔·盖茨与保罗·艾伦创始于 1975 年，总部设在华盛顿州的雷德蒙市。目前是全球最大的计算机软件提供商。其主要产品为 Windows 操作系统、Internet Explorer 浏览器、Microsoft Office 办公软件套件、SQL Server 数据库软件和开发工具等。1999 年推出了 MSN 网络即时信息客户程序，2001 年推出 Xbox 游戏机，参与游戏终端机市场竞争）借鉴 PHP 的思想，推出了 ASP 技术。ASP 脚本所使用的 VBScript 脚本语言直接来源于 VB 语言，它秉承了 VB 简单易用的特点，可以把脚本语言直接嵌入 HTML 文档中，无需编译连接就可运行，不存在浏览器兼容的问题。2002 年微软发布.NET 的正式版本.Net Framework 1.0，其中的 ASP 版本就是 ASP.NET 1.0，在此之前发布了两个.NET 测试版本 Beta1 和 Beta2。2003 年微软发布了.Net Framework 1.1 正式版，其中 ASP 版本就是 ASP.NET 1.1。2005 年微软发布了.Net Framework 2.0 正式版本，也就是 ASP.NET 2.0。微软在 2006 年发布了.Net Framework 3.0。

ASP.NET 目前能支持 3 种语言：C#、Visual Basic.NET 和 Jscript.NET。与 ASP 相比，ASP.NET 的优点如下：使用.NET 提供的所有类库，可以执行以往 ASP 所不能实现的许多功能；引入了服务器端控件的概念，这使开发交互式网站更加方便；引入了 ADO.NET 数据访问接口，大大提高了数据访问效率；提供 ASP.NET 的可视化开发环境 Visual studio.NET，进一步提高编程效率；保持对 ASP 的全面兼容；全面支持面向对象程序设计。

1997 年，Sun 公司推出了 Servlet 技术（Servlet 是运行在服务器端的程序，可以被认为是服务器端的 Applet。Servlet 被 Web 服务器（例如 Tomcat）加载和执行，就如同 Applet 被浏览器加载和执行一样。Servlet 从客户端（通过 Web 服务器）接收请求，执行某种操作，然后返回。它的主要优点包括持久性、平台无关性、可扩展性和安全性），成为 Java 阵营的 CGI 解决方案。1998 年，Sun 公司又推出了 JSP 技术，JSP 允许在 HTML 页面中嵌入 Java 脚本代码，从而实现动态网页功能。JSP 最大的优点是其开放的、跨平台的结构，它可以运行在所有的服务器系统上。2009 年 4 月 20 日，甲骨文（Oracle）公司以 74 亿美元收购 Sun 公司。

2000 年以后，随着 Web 应用程序复杂性不断提高，人们逐渐意识到，单纯依靠某种技术，很难实现快速开发、快速验证和快速部署的效果，必须整合 Web 开发技术形成完整的开发框架或应用模型，来满足各种复杂的应用程序的需求。Web 技术出现了几种主要的技术整合方式：MVC 设计模式、门户服务和 Web 内容管理。Struts、Spring、Hibernate 框架技术等都是开源世界里与 MVC 设计模式、门户服务和 Web 内容管理相关的优秀解决方案。

### 1.1.2 Web 动态网页技术

动态网页技术是运行在服务器端的 Web 应用程序，根据用户的请求，在服务器端进行动态处理后，把处理的结果以 HTML 文件格式返回给客户端。当前主流的三大动态 Web 开发技术是：PHP、ASP/ASP.NET 和 JSP。

**1. PHP**

1994 年 Rasmus Lerdorf 创建了 PHP。1995 年初 Personal Home Page Tools （PHP Tools）发布了 PHP 1.0；1995 年又发布 PHP 2.0；1997 年发布 PHP 3.0；2000 年，发布 PHP 4.0；2009 年发布 PHP 5.3 版本；2011 年发布 PHP 5.4。

PHP 是一个基于服务端来创建动态网站的脚本语言，可以用 PHP 和 HTML 生成网站主页。当一个访问者打开主页时，服务端便执行 PHP 的命令并将执行结果发送至访问者的浏览器中，这类似于 ASP 和 JSP。然而 PHP 和它们的不同之处在于 PHP 开放源码和跨越平台，PHP 可以运行在 Windows NT 和多种版本的 UNIX 上。PHP 消耗的资源较少，当 PHP 作为 Apache Web 服务器一部分时，运行代码不需要调用外部二进制程序，服务器不需要承担任何额外的负担。

2. ASP/ASP.NET

ASP（Active Server Pages，活动服务器页面）是一种允许用户将 HTML 或 XML 标记与 VBScript 代码或者 JavaScript 代码相结合生成动态页面的技术，用来创建服务器端功能强大的 Web 应用程序。当一个页面被访问时，VBScript/JavaScript 代码首先被服务器处理，然后将处理后得到的 HTML 代码发送给浏览器。ASP 只能建立在 Windows 的 IIS Web 服务器上。

ASP 是 Microsoft 公司开发、用于代替 CGI 脚本程序的一种 Web 应用技术，可以与数据库和其他程序进行交互，是一种简单、方便的编程工具。ASP 是基于 Web 的一种编程技术，是 CGI 的一种。ASP 可以轻松地实现对页面内容的动态控制，根据不同的浏览者，显示不同的页面内容。1996 年，Microsoft 公司推出 ASP 1.0；1998 年，Microsoft 公司推出 ASP 2.0；1999 年，Microsoft 公司推出 ASP 3.0；2001 年，推出 ASP.NET。

ASP.NET 技术又称为 ASP+，是在 ASP 基础上发展起来的，是 ASP 3.0 升级版本，保留 ASP 的最大优点并全力使其扩大化，是 Microsoft 公司推出的新一代 Web 开发技术，是.NET 战略中的重要一员，它全新的技术架构使编程变得更加简单，是创建动态网站和 Web 应用程序的最好技术之一。

3. JSP

JSP（Java Server Pages，Java 服务器页面）是由 Sun 公司倡导、许多公司参与共同建立的一种动态网页技术标准。JSP 技术类似 ASP/ASP.NET 技术，它是在传统的网页（HTML 文件）中插入 Java 代码段和 JSP 标记，从而形成 JSP 文件。Web 服务器接收到访问 JSP 网页的请求时，首先将 JSP 转化为 Servlet 文件，Servlet 文件经过编译后处理用户请求，然后将执行结果以 HTML 格式返回给客户。

1998 年，Sun 公司推出 JSP 0.9 版本；1999 年推出 JSP 1.1 版本；2000 年推出 JSP 1.2 版本。现在主要使用的是 JSP 2.0 版本。

自 JSP 推出后，许多大公司都支持 JSP 技术的服务器，如 IBM、甲骨文、微软公司等，所以 JSP 迅速成为主流商业应用的服务器端动态 Web 技术。

### 1.1.3 Web 应用程序的工作原理

JSP 页面是运行在服务器端的一种 Web 应用程序。在学习 JSP 技术前，先了解一下 Web 应用程序的工作原理。

目前，在 Internet 上的信息大多以网页形式存储在服务器上，通过浏览器获取网页内容，这是一种典型的 B/S（浏览器/服务器）模式。

B/S 模式的工作过程是：客户端请求-服务器处理-对客户端响应。

B/S 模式工作时，浏览器提交请求，Web 服务器接收到请求后把请求提交给相应的应用服务器，由应用服务器调用相应的 Web 应用程序对客户端请求进行处理，将处理结果返回给 Web 服务器，Web 服务器将处理结果（网页）响应给客户端（浏览器）。Web 应用程序由动态网页技术开发，如 JSP、ASP、PHP 等，其实现的原理如图 1-1 所示。

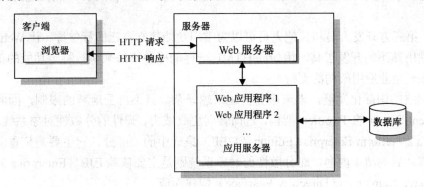

图 1-1　Web 应用程序的工作原理

## 1.2　JSP 技术概述

　　JSP 技术是由原 Sun 公司倡导、许多公司共同参与建立的一种基于 Java 语言的动态 Web 应用开发技术，利用这一技术可以建立安全、跨平台的先进动态网页技术。JSP 是 JavaEE 系统中的 Web 层技术，负责动态生成用户界面。JSP 页面在执行时采用编译方式，编译生成 Servlet 文件。

　　JSP 是从 Servlet 上分离出来的一小部分，简化了开发，加强了界面设计。JSP 定位在交互网页的开发。运用 Java 语法，但功能较 Servlet 弱了很多，并且高级开发中只充当用户界面部分。JSP 容器收到客户端发出的请求时，首先执行其中的程序片段，然后将执行结果以 HTML 格式响应给客户端。其中程序片段可以是：操作数据库、重新定向网页以及发送 E-Mail 等，这些都是建立动态网站所需要的功能。所有程序操作都在服务器端执行，网络上传送给客户端的仅是得到的结果，与客户端的浏览器无关，因此，JSP 被称为 Server-Side Language。JSP 的主要优点如下：

　　• 一次编写各处执行（Write once, Run Anywhere）

　　作为 Java 平台的一部分，JSP 技术拥有 Java 语言"一次编写，各处执行"的特点。随着越来越多的供货商将 JSP 技术添加到他们的产品中，用户可以针对自己公司的需求，做出谨慎评估后，选择符合公司成本及规模的服务器，假若未来的需求有所变更时，更换服务器平台并不影响之前所投下的成本、人力、所开发的应用程序。

　　• 搭配可重复使用的组件

　　JSP 技术可依赖于重复使用跨平台的组件（如：JavaBean 或 Enterprise JavaBean 组件）来执行更复杂的运算、数据处理。开发人员能够共享开发完成的组件，或者能够加强这些组件的功能，让更多用户或客户团体使用。基于善加利用组件的方法，可以加快整体开发过程，也大大降低公司的开发成本和人力。

　　• 采用标签化页面开发

　　Web 网页开发人员不一定都是熟悉 Java 语言的程序员。因此，JSP 技术能够将许多功能封装起来，成为一个自定义的标签，这些功能是完全根据 XML 的标准来制订的，即 JSP 技术中的标签库（Tag Library）。因此，Web 页面开发人员可以运用自定义好的标签来达成工作需求，而无须再写复杂的 Java 语句，让 Web 页面开发人员亦能快速开发出动态内容网页。

今后，第三方开发人员和其他人员可以为常用功能建立自己的标签库，让 Web 网页开发人员能够使用熟悉的开发工具，用如同 HTML 一样的标签语法来执行特定功能的工作。

• N-tier 企业应用架构的支持

鉴于网络的国际化发展，未来服务会越来越繁杂，且不再受地域的限制，因此，必须放弃以往 Client-Server 的 Two-tier 架构，进而转向更具威力、弹性的分散性对象系统。由于 JSP 技术是 Java 2 Platform Enterprise Edition（J2EE）集成中的一部分，它主要是负责前端显示经过复杂运算之后的结果内容，而分散性的对象系统则是主要依赖 EJB（Enterprise JavaBean）和 JNDI（Java Naming and Directory Interface）构建而成。

## 1.3 JSP 的工作原理

JSP 应用程序是运行在服务器端。服务器端收到用户通过浏览器提交的请求后进行处理，再以 HTML 的形式返回给客户端，客户端得到的只是在浏览器中看到的静态网页。JSP 工作原理如图 1-2 所示。

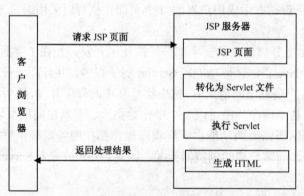

图 1-2  JSP 工作原理

所有的 JSP 应用程序在首次载入时都被编译成 Servlet 文件，然后再运行，这个工作主要是由 JSP 引擎来完成。当第一次运行一个 JSP 页面时，JSP 引擎要完成以下操作：

（1）当用户访问一个 JSP 页面时，JSP 页面将被编译成 Servlet 文件（Java 文件）。
（2）JSP 引擎调用 Java 编译器，编译 Servlet 文件为可执行的代码文件（.class 文件）。
（3）用 Java 虚拟机（JVM）解释执行.class 文件，并将执行结果返回给服务器。
（4）服务器将执行结果以 HTML 格式发送给客户端的浏览器。

由于一个 JSP 页面在第一次被访问时要经过编译成 Servlet 文件、Servlet 编译和.class 文件执行这几个步骤，所以客户端得到响应所需要的时间比较长。当该页面再次被访问时，它对应的.class 文件已经生成，不需要再次翻译和编译，JSP 引擎可以直接执行.class 文件，因此 JSP 页面的访问速度会大大提高。

## 1.4 JSP 程序开发模式

### 1.4.1 JSP 两种体系结构

Sun 公司早期提出了两种使用 JSP 技术建立 Java Web 应用程序的方式。

**1. JSP Model 1**

在 Model 1 体系中，JSP 页面独自响应请求并将处理结果返回客户，如图 1-3 所示。这里仍然存在显示与内容的分离，因为所有的数据存取都是由 JavaBean 来完成的。尽管 Model 1 体系十分适合简单应用的需要，它却不能满足复杂的大型 Java Web 应用程序需要。不加选择地随意运用 Model 1，会导致 JSP 页面内被嵌入大量的脚本片段或 Java 代码。尽管这对于 Java 程序员来说可能不是什么大问题，但如果 JSP 页面是由网页设计人员开发并维护的，这就确实是个问题了。从根本上讲，将导致角色定义不清和职责分配不明，给项目管理带来不必要的麻烦。

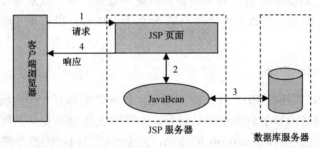

图 1-3　JSP Model 1 模型结构图

**2. JSP Model 2**

Model 2 体系结构是一种把 JSP 与 Servlet 联合使用来实现动态内容服务的方法，如图 1-4 所示。它吸取了两种技术各自的优点，用 JSP 生成表示层（view）的内容，让 Servlet 完成深层次的处理任务。Servlet 充当控制者（Controller）的角色，负责管理对请求的处理，创建 JSP 页面需要使用的 JavaBean 和对象，同时根据用户的动作决定把哪个 JSP 页面传给请求者。在 JSP 页面内没有处理逻辑，它仅负责检索原先由 Servlet 创建的对象或者 JavaBean，从 Servlet 中提取动态内容插入静态模板。这种格式分离了显示和内容，明确了角色的定义以及实现了开发者与网页设计者的分开，项目越复杂，使用 Model 2 体系结构的优势就越突出。

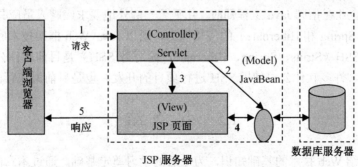

图 1-4　JSP Model 2 模型结构图

### 1.4.2 JSP 开发 Java Web 站点的主要方式

JSP 是 JavaEE 的一部分,可以用于开发小型的 Web 站点,也可以用于开发大型的、企业级的应用程序。根据开发的目标程序不同,使用的开发方式也不同。JSP 开发 Web 站点主要有以下几种方式。

**1. 直接使用 JSP**

对于最小型的 Web 站点,可以直接使用 JSP 来构建动态网页,这种站点最为简单,开发简单应用程序,如简单的留言板、动态日期。对于这种开发模式,一般可以将所有的动态处理部分都放置在 JSP 文件中。

**2. JSP+JavaBean**

中型站点面对的是数据库查询、用户管理和小量的商业业务逻辑。对于这种站点,不能将所有的数据全部交给 JSP 页面来处理。在单纯的 JSP 中加入 JavaBean 技术将有助于这种中型站点的开发。利用 JavaBean 将很容易对诸如数据库连接、用户登录与注册、商业业务逻辑等进行封装。例如,将常用的数据库连接写成一个 JavaBean,既方便了使用,又可以使 JSP 文件简单而清晰。

**3. JSP +Servlet+JavaBean**

无论使用 ASP.NET 还是 PHP 开发动态网站,长期以来都有一个比较重要的问题,就是网站的逻辑关系和网站的显示页面不容易分开。在逻辑关系异常复杂的网站中,借助于 JSP 和 Servlet 良好的交互关系和 JavaBean 的协助,完全可以将网站的整个逻辑结构放在 Servlet 中,而将动态页面的输出放在 JSP 页面中来完成。在这种开发方式中,一个网站可以由一个或几个核心的 Servlet 来处理网站的逻辑,通过调用 JSP 页面来完成客户端的请求。

**4. JavaEE 开发模型**

在 JavaEE 开发模型中,整个系统可以分为 3 个主要的部分:视图、控制器和模型。视图就是用户界面部分,主要处理用户看到的界面;控制器负责网站的整体逻辑,用于管理用户与视图发生的交互;模型是应用业务逻辑部分,主要由 EJB 负责完成,借助于 EJB 强大的组件技术和企业级的管理控制,开发人员可以轻松地创建出可重用的业务逻辑模块。

**5. 框架整合应用**

目前,软件企业在招聘 Java 工程师时,几乎无一例外地要求应聘人员应具备 Java Web 框架技术(Struts、Spring 和 Hibernate)的应用能力,所以 Java Web 框架技术应用是 Java 工程师必备的技能。SSH(Struts、Spring、Hibernate,简写为 SSH)是目前软件公司用到的 3 个主流的开源框架,许多软件公司使用 SSH 进行项目的开发,也是目前最流行的开发模式。

## 1.5 小 结

本章主要介绍 Web 技术的基础知识,为今后的学习奠定基础。通过本章的学习,应该掌握以下内容:

- Web 应用程序。
- JSP 技术概述。
- JSP 工作原理。
- JSP 程序开发模式。

总之，本章是进行 Java Web 开发以及后续章节学习的基础，通过本章学习，应理解 Web 技术的基本知识。

## 1.6 习　　题

1. 简述 JSP 的工作原理。
2. 简述 JSP 两种体系结构。
3. 简述 JSP 开发 Web 站点的主要方式。

# 第 2 章  JSP 开发环境搭建

## 2.1  JSP 的运行环境

本章主要讲解 JSP 开发环境的搭建方法、集成开发工具 Eclipse 的安装和配置，掌握使用开发工具创建 Java Web 应用，并简要介绍 BBS 项目案例的系统功能和数据库设计。

### 2.1.1  JSP 的运行环境

使用 JSP 进行开发，需要具备以下对应的运行环境：Web 浏览器、Web 服务器、JDK 开发工具包以及数据库。下面分别介绍这些环境。

1．Web 浏览器

浏览器是用于客户端用户访问 Web 应用的工具，与开发 JSP 应用不存在很大的关系，所以开发 JSP 对浏览器的要求并不是很高，任何支持 HTML 的浏览器都可以。

2．Web 服务器

JSP 运行需要在计算机上安装 JSP 引擎，Java Web 应用程序需要部署运行在 Java Web 服务器中，常用的 Java Web 服务器有 Tomcat、GlassFish、WebLogic、JBoss、WebSphere、Jetty、JRun 等。

Tomcat 是 Apache 软件基金会（Apache Software Foundation）的 Jakarta 项目中的一个核心项目，由 Apache、Sun 和其他一些公司及个人共同开发而成。由于有了 Sun 的参与和支持，最新的 Servlet 和 JSP 规范总是能在 Tomcat 中得到体现，Tomcat 5 支持最新的 Servlet 2.4 和 JSP 2.0 规范。因为 Tomcat 技术先进、性能稳定，而且免费，因而深受 Java 爱好者的喜爱，并得到了部分软件开发商的认可，成为目前比较流行的 Web 应用服务器。目前最新版本是 7.0。

GlassFish 是一款强健的商业兼容应用服务器，达到产品级质量，可免费用于开发、部署和重新分发。GlassFish 用于构建 Java EE 5 应用服务器的开源开发项目的名称。它基于 Sun Microsystems 提供的 Sun Java System Application Server PE 9 的源代码以及 Oracle 贡献的 TopLink 持久性代码。该项目提供了开发高质量应用服务器的结构化过程，以前所未有的速度提供新的功能。这是对希望能够获得源代码并为开发 Sun 的下一代应用服务器（基于 GlassFish）作出贡献的 Java 开发者作出的回应。该项目旨在促进 Sun 和 Oracle 工程师与社区之间的交流，它将使得所有开发者都能够参与到应用服务器的开发过程中来。

WebLogic 是美国 BEA 公司出品的一个 application server，确切地说是一个基于 Java EE 架构的中间件，BEA WebLogic 是用于开发、集成、部署和管理大型分布式 Web 应用、网络应用和数据库应用的 Java 应用服务器，它将 Java 的动态功能和 Java Enterprise 标准的安全性引入大型网络应用的开发、集成、部署和管理之中。

JBoss 是全世界开发者共同努力的成果，一个基于 J2EE 的开放源代码的应用服务器。因为 JBoss 代码遵循 LGPL 许可，可以在任何商业应用中免费使用它，而不用支付费用。2006 年，Jboss 公司被 Redhat 公司收购。JBoss 是一个管理 EJB 的容器和服务器，支持 EJB 1.1、EJB 2.0 和 EJB3.0 的规范。但 JBoss 核心服务不包括支持 servlet/JSP 的 WEB 容器，一般与 Tomcat 或 Jetty 绑定使用。

WebSphere 是 IBM 的软件平台。它包含了编写、运行和监视全天候的工业强度的随需应变 Web 应用程序和跨平台、跨产品解决方案所需要的整个中间件基础设施，如服务器、服务和工具。WebSphere 提供了可靠、灵活和健壮的软件。

Jetty 是一个开源的 Servlet 容器，它为基于 Java 的 Web 内容，例如 JSP 和 Servlet 提供运行环境。Jetty 是使用 Java 语言编写的，它的 API 以一组 JAR 包的形式发布。开发人员可以将 Jetty 容器实例化成一个对象，可以迅速为一些独立运行（stand-alone）的 Java 应用提供网络和 Web 连接。

3．JDK

JDK（Java Development Kit）是 Sun 公司（已被 Oracle 收购）针对 Java 开发人员的软件开发工具包。自从 Java 推出以来，JDK 已经成为使用最广泛的 Java SDK（Software development kit）。JDK 是整个 Java 的核心，包括了 Java 运行环境、Java 工具和 Java 基础类库。作为 Java 语言的 SDK，普通用户并不需要安装 JDK 来运行 Java 程序，而只需要安装 JRE（Java Runtime Environment）。而程序开发者必须安装 JDK 来编译、调试程序。

4．数据库

数据库是数据的结构化集合，是按照数据结构来组织、存储和管理数据的仓库。数据可以是任何东西，从简单的购物清单到图像，或企业网络中的海量信息。要想将数据添加到数据库，或访问、处理计算机数据库中保存的数据，需要使用数据库管理系统，如 MySQL 服务器。目前 MySQL 被广泛地应用在 Internet 上的中小型网站中。由于其体积小、速度快、总体拥有成本低，尤其是开放源码这一特点，许多中小型网站为了降低网站总体拥有成本，而选择了 MySQL 作为网站数据库。Java 数据库连接（JDBC）由一组用 Java 编程语言编写的类和接口组成。JDBC 为工具、数据库开发人员提供了一个标准的 API，使他们能够用纯 Java API 来编写数据库应用程序。

## 2.1.2 JDK 的安装与配置

JDK 由 Oracle 公司提供，其中包括运行 Java 程序所必需的 JRE（Java 运行时环境）及开发过程中常用的库文件。在使用 JSP 开发 Web 应用之前，首先必须安装 JDK 组件。

1．JDK 下载与安装

JDK 是一个开源、免费的工具。可以到 Oracle 公司的官方网站上下载 JDK 最新版本，网址为：http://www.oracle.com/technetwork/java/javase/downloads/jdk7-downloads-1880260.html，如图 2-1 所示。

下载 JDK 需要选择和自己机器操作系统匹配的版本，如图 2-2 所示。微软的操作系统有 32 位（x86）和 64 位（x64）之分，x64 的 CPU 可以用 x64 的软件，也可以用 x86 的软件，而 x86 的 CPU 电脑只能用 x86 的软件。如果计算机的操作系统是 Windows x64，则应选择 JDK 版本是支持 Windows x64（90.4 MB）的 jdk-7u15-windows-x64.exe 下载。

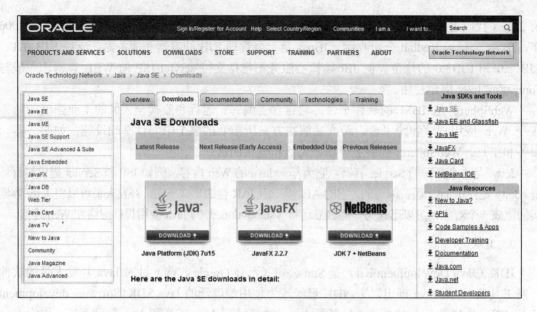

图 2-1　JDK 下载页面

**Java SE Development Kit 7u15**

You must accept the Oracle Binary Code License Agreement for Java SE to download this software.

Thank you for accepting the Oracle Binary Code License Agreement for Java SE; you may now download this software.

| Product / File Description | File Size | Download |
|---|---|---|
| Linux x86 | 106.64 MB | jdk-7u15-linux-i586.rpm |
| Linux x86 | 92.97 MB | jdk-7u15-linux-i586.tar.gz |
| Linux x64 | 104.77 MB | jdk-7u15-linux-x64.rpm |
| Linux x64 | 91.68 MB | jdk-7u15-linux-x64.tar.gz |
| Mac OS X x64 | 143.75 MB | jdk-7u15-macosx-x64.dmg |
| Solaris x86 (SVR4 package) | 135.52 MB | jdk-7u15-solaris-i586.tar.Z |
| Solaris x86 | 91.94 MB | jdk-7u15-solaris-i586.tar.gz |
| Solaris SPARC (SVR4 package) | 135.92 MB | jdk-7u15-solaris-sparc.tar.Z |
| Solaris SPARC | 95.26 MB | jdk-7u15-solaris-sparc.tar.gz |
| Solaris SPARC 64-bit (SVR4 package) | 22.92 MB | jdk-7u15-solaris-sparcv9.tar.Z |
| Solaris SPARC 64-bit | 17.59 MB | jdk-7u15-solaris-sparcv9.tar.gz |
| Solaris x64 (SVR4 package) | 22.53 MB | jdk-7u15-solaris-x64.tar.Z |
| Solaris x64 | 14.96 MB | jdk-7u15-solaris-x64.tar.gz |
| Windows x86 | 88.75 MB | jdk-7u15-windows-i586.exe |
| Windows x64 | 90.4 MB | jdk-7u15-windows-x64.exe |

图 2-2　选择与操作系统相适应的 JDK 版本

本次安装选择的是 jdk-7u15-windows-i586.exe，双击 JDK 可执行文件开始在线安装。根据安装向导依次执行安装，如图 2-3 所示。

安装过程中可以选择更改 JDK 的默认安装路径，如图 2-4 所示。安装好的 JDK 目录结构如图 2-5 所示。

bin 目录：JDK 开发工具的可执行文件。

lib 目录：开发工具使用的归档包文件。

jre 目录：Java 运行时环境的根目录，包含 Java 虚拟机，运行时的类包和 Java 应用启动器，但不包含开发环境中的开发工具，

include 目录：包含 C 语言头文件，支持 Java 本地接口与 Java 虚拟机调试程序接口的本地编程技术。

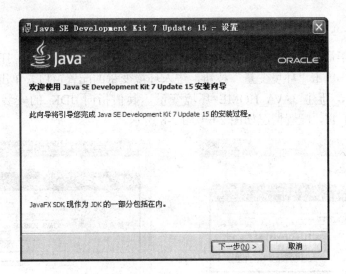

图 2-3　JDK 安装向导

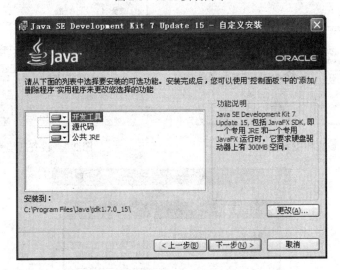

图 2-4　选择更改安装路径

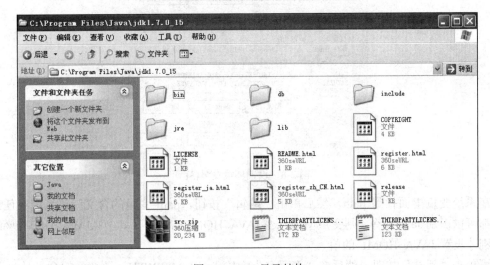

图 2-5　JDK 目录结构

2. JDK 的配置与测试

JDK 安装完毕后，需要配置操作系统的环境变量。右击"我的电脑"，打开"属性"窗口，选择"高级"栏，单击"环境变量"按钮，进入环境变量的配置窗口。单击"系统变量"中的"新建"按钮，新建 JAVA_HOME 环境变量，其值指向 JDK 的安装目录 C:\Program Files\Java\jdk1.7.0_15，配置情况如图 2-6～图 2-8 所示。

图 2-6　设置环境变量(1)　　　　　　　　图 2-7　设置环境变量(2)

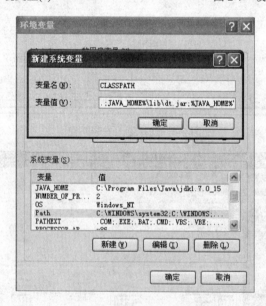

图 2-8　设置环境变量(3)

在系统变量里面找到 path，然后单击"编辑"按钮。path 变量的含义就是系统在任何路径下都可以识别 Java 命令，则变量值为".;%JAVA_HOME%\bin"（其中%JAVA_HOME%的意思为刚才设置 JAVA_HOME 的值）。

再单击"新建"按钮，然后在"变量名"中输入 CLASSPATH。该变量的含义是为 Java

加载类（class or lib）路径。Java 进行编译、运行时需要用到的 jar 包位置，只有类在 CLASSPATH 中，Java 命令才能识别。其值为 ".;JAVA_HOME%\lib\dt.jar;%JAVA_HOME%\lib\toos.jar"，要加 "." 表示当前路径。

以上 3 个变量设置完毕，则按 "确定" 按钮直至属性窗口消失。按下来是验证安装是否成功。先打开 "开始" -> "运行"，输入 cmd，进入 DOS 系统界面。然后输入 java -version，如果安装成功，系统会显示 java version "1.7.0_15" 等 JDK 信息，如图 2-9 所示。

图 2-9　JDK 安装测试

### 2.1.3　Tomcat 的安装与启动

#### 1. Tomcat 的安装

作为一个开源的 Web 服务器，Tomcat 提供了简易的安装，如图 2-10 所示。首先需要在网站上下载 Tomcat 安装文件[版本号].[后缀名]，安装文件有 zip 格式、tar.gz 格式，也有 exe 格式。对于 zip 格式的安装文件，只需要将其解压缩到某目录下即可；tar.gz 格式文件适用于 UNIX 及其相关操作系统；而 exe 格式的图形化安装文件则更加符合 Windows 用户的操作习惯。Tomcat 的下载地址为：http://tomcat.apache.org/download-70.cgi，其下载页面如图 2-10 所示。

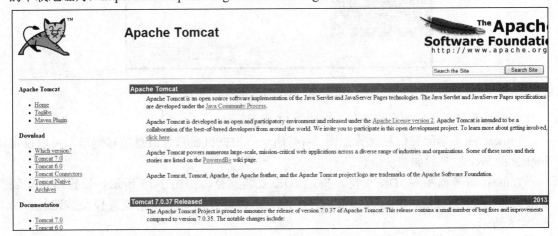

图 2-10　Tomcat 下载页面

Tomcat 服务器安装也需要选择与机器相适应的版本，如图 2-11 所示。

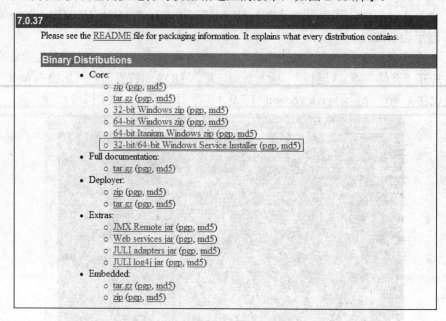

图 2-11　Tomcat 版本选择

Tomcat 安装向导如图 2-12 所示，安装过程中根据提示依次选择安装选项。

图 2-12　Tomcat 安装向导

Tomcat 服务器组件的安装选择，如图 2-13 所示。

Tomcat 服务器配置信息设置如图 2-14 所示，根据机器系统信息填写端口，可以设定 Tomcat 的 Web 管理界面登录信息。

Tomcat 服务器需要 JDK 支持，指定 JDK 安装路径，如图 2-15 所示。设置默认安装安装路径为 C:\Program Files\Java\jdk1.7.0_15，可以通过浏览文件系统选择安装路径，如图 2-16 所示。

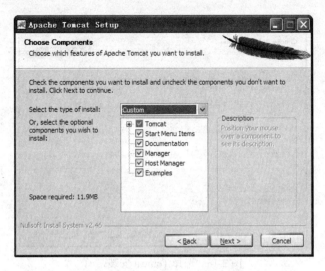

图 2-13  选择安装 Tomcat 组件

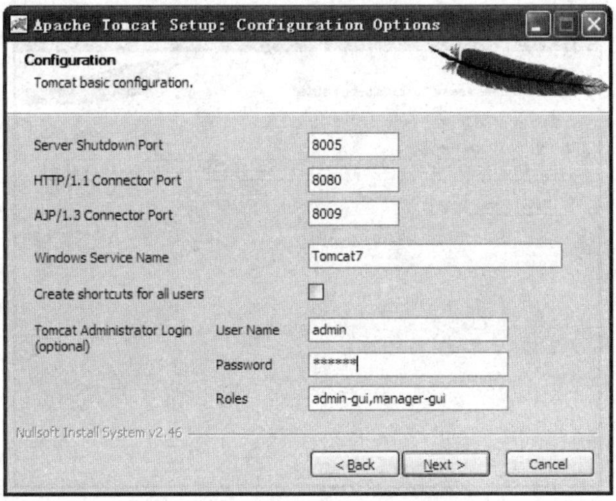

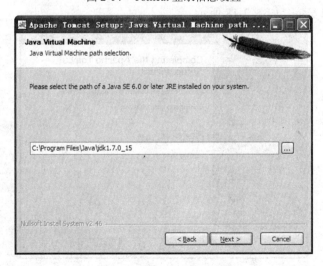

图 2-14  Tomcat 登录信息设置

图 2-15  选择 Java 虚拟机路径

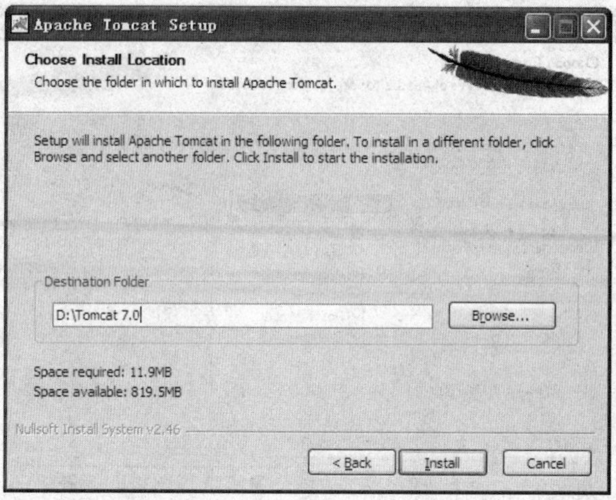

图 2-16 选择 Tomcat 安装路径

开始安装 Tomcat 7.0，如图 2-17 所示。

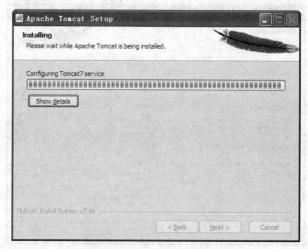

图 2-17 Tomcat 安装界面

完成 Tomcat 的安装，如图 2-18 所示。

图 2-18 Tomcat 安装成功

## 2. Tomcat 的启动

在系统程序菜单中启动 Tomcat，如图 2-19 所示。打开 Tomcat 管理窗口，单击 Start 按钮启动，如图 2-20 所示。

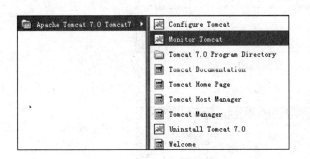

图 2-19　Tomcat 启动菜单　　　　　　　图 2-20　Tomcat 启动

### 2.1.4　JSP 开发工具——Eclipse

Eclipse 是一个基于 Java 的、开放源码的、可扩展的应用开发平台，它为编程人员提供 Java 集成开发环境。它是一个可以用于构建集成 Web 和应用程序开发工具的平台。

#### 1. 下载解压 Eclipse 软件

JSP 开发工具 Eclipse 的下载地址为http://www.eclipse.org/downloads/。Eolipse 的下载界面如图 2-21 所示。

下载后解压到本地文件夹中。

图 2-21　新建动态 Web 项目

#### 2. Eclipse-Java EE 下创建 Tomcat server

启动 Eclipse 软件，在 Java EE-Eclipse 视图下创建 Tomcat Server。当前透视图是 Java EE，如图 2-22 所示。在 Eclipse 窗体的右下部分有多个选项卡。

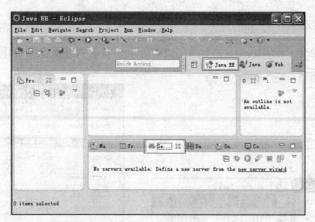

图 2-22　Eclipse 的 Java EE 透视图

在 Servers 面板上右击，创建新的 Server，如图 2-23 所示。

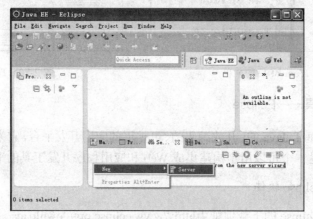

图 2-23　配置 Eclipse 的服务器

选择已经安装的 Tomcat 版本，如图 2-24 所示。

图 2-24　在 Eclipse 中选择 Tomcat 7 服务器

指定 Tomcat 的安装路径，如图 2-25 所示。

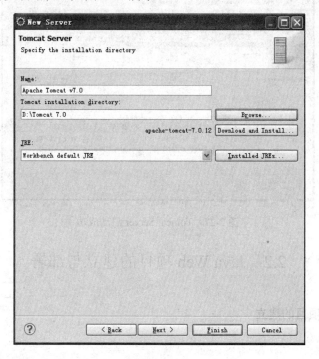

图 2-25　指定 Tomcat 安装路径

3. 启动 Eclipse 中的 Tomcat Server

测试在 Eclipse-Java EE 下创建的 Tomcat Server 是否成功，在 Servers 面板上右击，执行 start，或者单击工具栏上的绿色"运行"按钮启动 Tomcat Server，如图 2-26 所示。

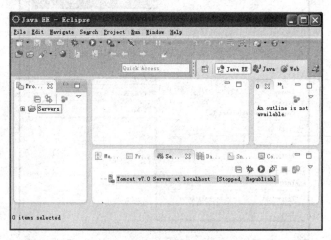

图 2-26　Tomcat server 启动状态

Eclipse 中的 Tomcat Server 状态由原来[Stopped,Republish]变为[Started,Synchronized]，说明启动成功，如图 2-27 所示。

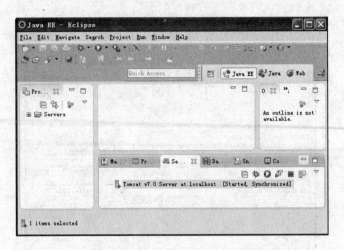

图 2-27 Tomcat Server 启动成功

## 2.2 Java Web 项目的建立与部署

### 2.2.1 Java Web 项目的建立

1. 新建项目

在 Eclipse 平台界面的左侧的 Project Explorer 右击，选择 New -> Project，或在 File 菜单中选择 New->Project，如图 2-28 所示。弹出 New Project 窗口。

图 2-28 新建一个项目

2. 创建动态 Web 项目

选择 Dynamic Web Project，如图 2-29 所示。

图 2-29　选择动态 Web 项目

然后单击 Next 按钮，在 Project Name 中输入想要的项目名称。输入后单击 Finish 按钮完成新项目的创建，如图 2-30 所示。

图 2-30　输入 Web 项目名称

3. 配置项目的运行时环境

配置项目的运行时环境如图 2-30 所示。

4. 选择 Tomcat 服务器安装位置

在运行时环境界面中选择 Tomcat 服务器安装路径，如图 2-31 所示。

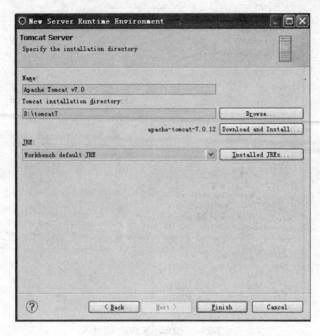

图 2-31 选择 Tomcat 安装路径

5. 动态 Web 项目的运行时环境配置

在动态 Web 项目界面中选择 Tomcat 作为运行时环境，如图 2-32 所示。

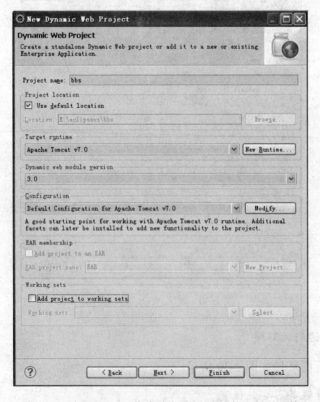

图 2-32 选择 Tomcat 作为动态 Web 项目的运行时环境

6. 动态 Web 项目的上下文路径

在动态 Web 项目界面中配置上下文路径，如图 2-33 所示。

图 2-33　配置 Web 项目的上下文路径

7. 动态 Web 项目的文件结构

安装好的动态 Web 项目的文件结构如图 2-34 所示。

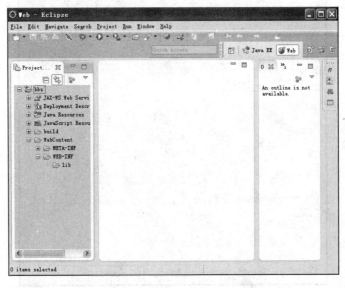

图 2-34　动态 Web 项目的文件结构

## 2.2.2　创建一个 JSP

在 bbs 项目名上右击，选择 New -> JSP File，如图 2-35 所示。在弹出的 New JSP File 窗口中输入 File name：register.jsp，单击 Finish 按钮，如图 2-36 所示。

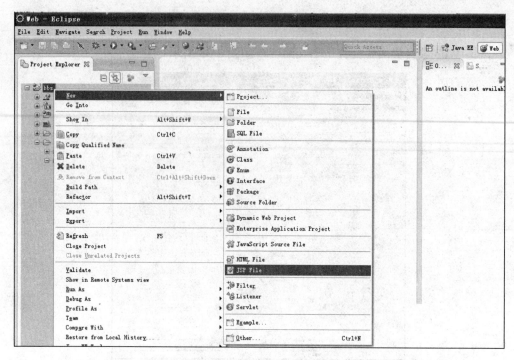

图 2-35　动态 Web 项目文件结构

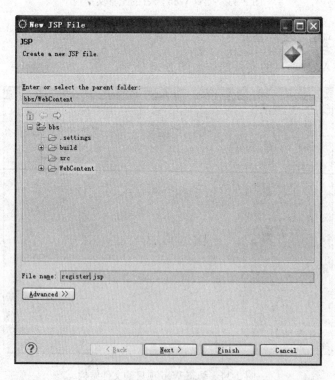

图 2-36　输入 JSP 文件名

在 New JSP File 界面中选择使用 JSP 模板，如图 2-37 所示。

打开新创建的 JSP 文件，界面右侧显示文件内容，如图 2-38 所示。

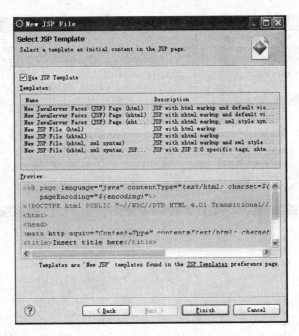

图 2-37 使用 JSP 模板

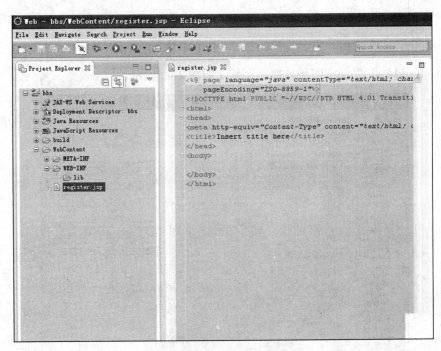

图 2-38 JSP 文件内容

## 2.2.3 创建 Servlet

### 1. 项目添加库文件

创建 Servlet 类前需要先导入一个 jar 包 Servlet-api.jar，这是 Servlet 的 jar 包。在 Tomcat 的 lib 文件下，有个 servlet-api.jar 包。在 bbs 项目名上右击，选择 Build Path->Configure Build Path，如图 2-39 所示。

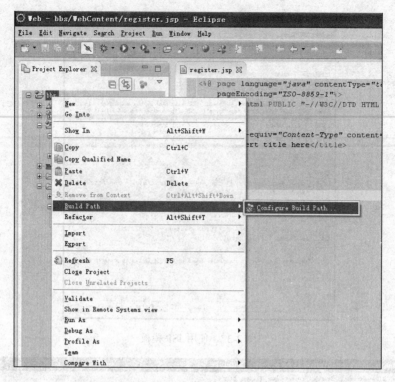

图 2-39 配置构建路径

在 Java Build Path 界面中添加外部库文件,如图 2-40 所示。

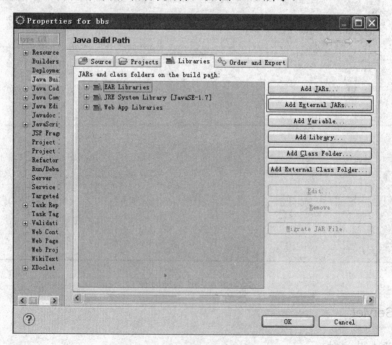

图 2-40 添加外部库文件

然后定位到 Tomcat 的 Lib 文件夹,选择 servlet-api.jar,如图 2-41 所示。

项目 Libraries 库文件结构如图 2-42 所示。

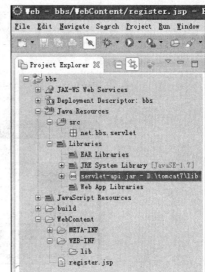

图 2-41　添加 servlet-api.jar 库文件　　　　图 2-42　项目 Libraries 库文件结构

2. 创建 Servlet 类

在 bbs 项目名上右击，选择 New -> Class，如图 2-43 所示。

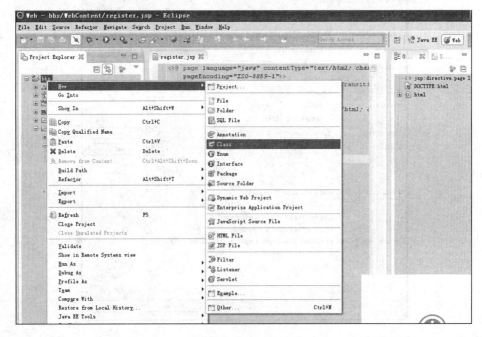

图 2-43　创建类文件

在弹出的 New Java Class 界面中输入包名和类名，并声明其继承父类 Http Servlet，如图 2-44 所示。

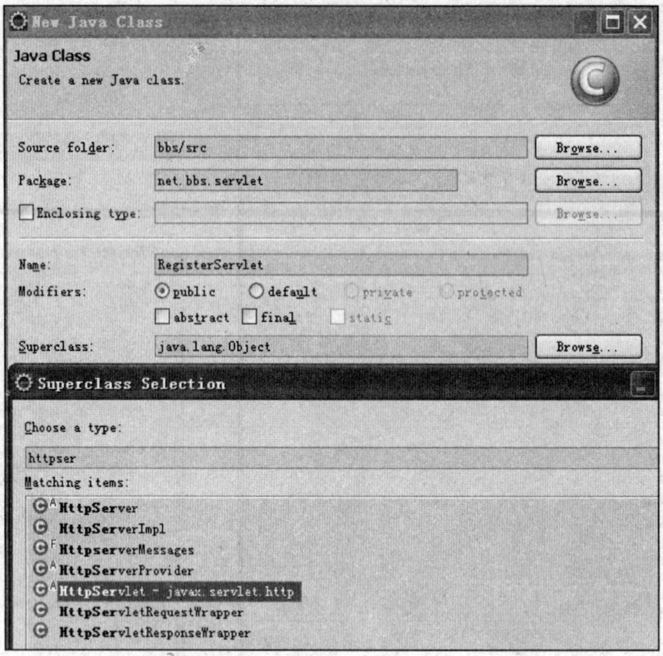

图 2-44  定义 servlet 类

Servlet 类信息如图 2-45 所示。单击 Finish 按钮。

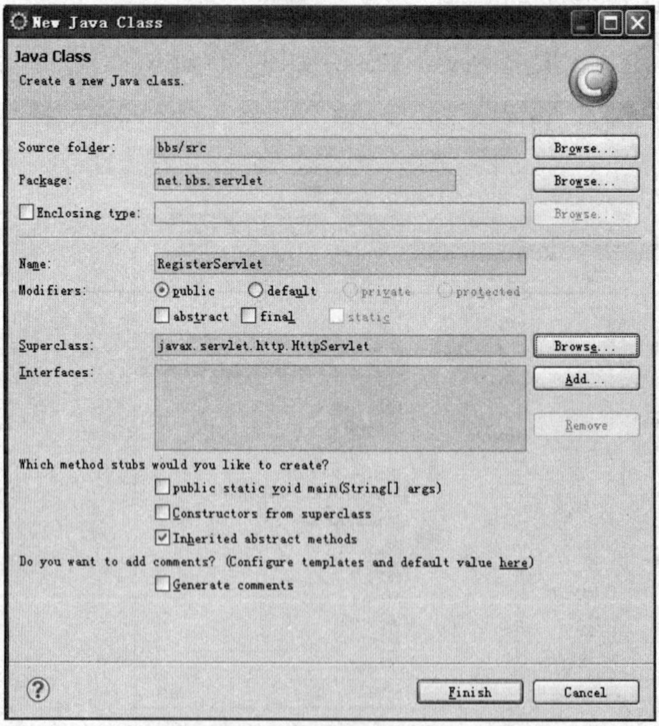

图 2-45  Servlet 类信息

编写 Servlet 类，如图 2-46 所示。

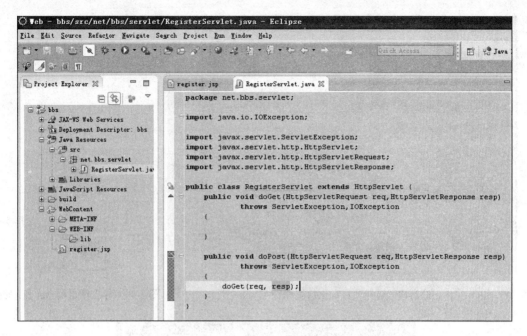

图 2-46　编写 Servlet 类

## 2.2.4　Java Web 项目的部署

**1. 导出 Java Web 项目**

在 bbs 上右击，选择 Export→WAR file，导出 Java Web 项目，如图 2-47 所示。

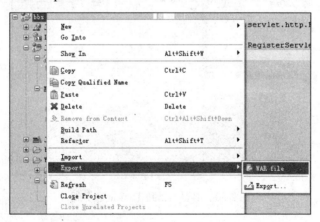

图 2-47　导出 Java Web 项目

在 Export 界面中选项项目导出放置的路径，如图 2-48 所示。

导出项目的 war 压缩包如图 2-49 所示。

**2. 部署项目**

复制 war 文件到 Tomcat 的 webapps 文件夹中，如图 2-50 所示。启动 Tomcat 服务器，会自动解压 Web 项目并运行。

图 2-48 选择项目导出放置路径　　　　　　　　图 2-49 导出项目的 war 压缩包

图 2-50 部署项目到 Tomcat 服务器

## 2.3 BBS 案例的分析与设计

本文以 BBS 项目的内容实现作为学习 JSP 编程知识点案例，通过系统地串讲 BBS 项目案例，熟悉 JSP 项目设计过程，掌握 Java Web 开发技术。

### 2.3.1 系统分析

系统分析的目的是为系统设计通过系统的逻辑模型，系统设计再根据逻辑模型进行物理方案的设计。

1. 需求描述

论坛也称为 BBS，是 Bulletin Board System 的简称，即电子公告板，是一种在 Internet 上常见的、用于信息服务的 Web 系统，它主要给浏览者提供相互沟通的平台，以此来吸引用户，服务用户。

论坛系统的主要目的是为所有访问该论坛的用户提供一个可以方便、自由地发表观点的网上场所，因此需要为用户提供一个简捷、方便的操作界面，同时也为具有高级权限的管理员提供相应的维护功能。要求实现以下功能：浏览帖子、发表新帖、回复帖子、首页分页显示主题帖子简要信息列表，统计每条帖子的点击量和回复次数等。同时为管理员设置高级权限，使他可以对不健康的主题帖子和回复信息进行删除。在做删除处理时，如果删除的是主题帖，应同时删除其对应的所有回复信息；如果删除的是回复信息，要求其回复次数同时减 1。

2. 用例分析

用例图显示外部参与者与系统的交互，能够更直观地描述系统的功能。从角色来看，论坛系统主要涉及 4 个角色：游客（未登录）、会员（登录用户）、版主、管理员。论坛系统的用例图如图 2-51 所示。

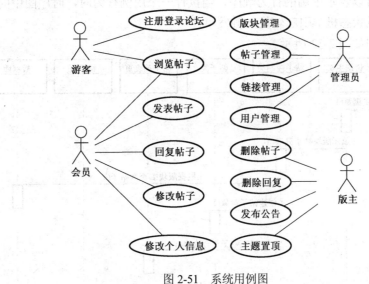

图 2-51　系统用例图

### 2.3.2　系统设计

1. 模块功能划分

按照系统角色的使用的系统功能划分为 4 个模块，如表 2-1 所示。

表 2-1　bbs_userinfo

| 模块名称 | 功能名称 | 功能描述 |
| --- | --- | --- |
| 用户管理模块 | 用户注册 | 用户输入自己的基本信息，系统验证信息的有效性，并将用户信息存入数据库 |
| | 用户登录 | 用户输入自己的用户名和密码，系统检验用户名和密码的有效性，对合法用户基于角色授权 |
| | 用户信息修改 | 用户可以对自己的基本信息进行修改 |
| | 删除用户 | 管理员可以删除那些不符合 BBS 管理规则，长时间不登录的用户 |

续表

| 模块名称 | 功能名称 | 功能描述 |
| --- | --- | --- |
| 版块管理模块 | 浏览版块 | 用户（包括游客）可以浏览论坛中的各版块 |
| | 添加版块 | 管理员可以根据用户的需要添加新的版块 |
| | 删除版块 | 管理员可以对不能为用户提供更好信息的版块进行删除 |
| | 指定版主 | 管理员可以根据需要设置任意注册用户为任意版块的版主 |
| 帖子管理模块 | 浏览帖子 | 用户（包括游客）可以浏览论坛中的帖子 |
| | 发表帖子 | 注册用户可以在自己感兴趣的版块中发表新的帖子 |
| | 删除帖子 | 管理员可以对要进行删除的所有帖子进行删除，版主只能对自己版块内的帖子进行删除 |
| | 回复帖子 | 注册用户可以对自己感兴趣的主题回复帖子 |
| | 检索帖子 | 用户（包括游客）可以通过输入关键字检索自己感兴趣的帖子 |
| | 指定精华 | 当管理员发现用户所发表的帖子非常好时将该帖指定为精华，版主只能对自己版块内的帖子指定为精华 |
| | 帖子置顶 | 管理员可以对点击率高和回复率高的帖子进行置顶，版主只能对自己版块内的帖子进行置顶 |
| 系统公告、链接管理模块 | 链接管理 | 管理员可以设置著名BBS网站的链接，可以对链接进行添加、修改、删除操作 |
| | 公告管理 | 管理员可以发布系统公告，可以对系统公告进行添加、修改、删除操作 |

### 2. 时序图设计

时序图描述系统的动态行为，通过描述对象之间发送消息的时间顺序，显示多个对象之间的动态协作。它可以表示用例的行为顺序，当执行一个用例行为时，时序图中的每条消息对应了一个类操作或状态机中引起转换的触发事件。

会员发帖时序图如图2-52所示。

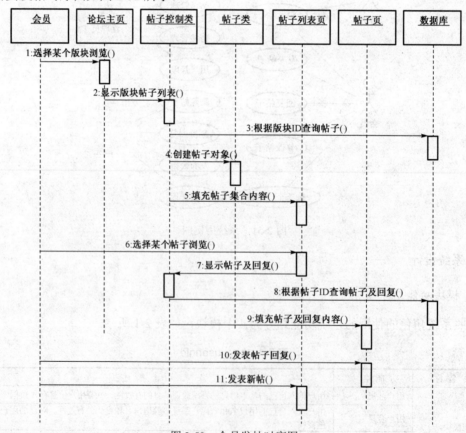

图2-52 会员发帖时序图

管理员添加版块时序图如图2-53所示。

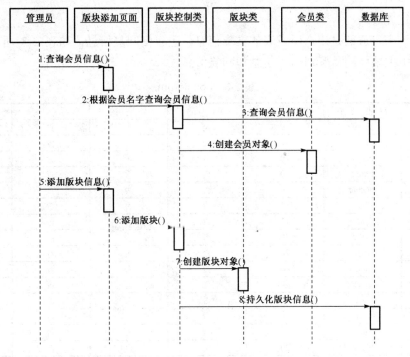

图 2-53 管理员添加版块时序图

3. 数据库设计

系统数据库命名为 bbsdb，6 个数据表分别为用户信息表（bbs_userinfo）、管理员表（bbs_admin）、论坛文章表（bbs_forum）、论坛版块表（bbs_board）、BBS 网站链接表（bbs_link）、站内公告表（bbs_notice）。

（1）用户信息表

用户信息表中存放的是用户登录时所需要或者记录用户的基本信息，包含比如登录名、密码、最后登录时间等。用户基本信息表的建立主要是记录用户最常用的一些信息，如表 2-2 所示。

表 2-2  bbs_userinfo

| 字段名 | 数据类型 | 长度 | 主键 | 外键 | 可空 | 说明 |
| --- | --- | --- | --- | --- | --- | --- |
| ID | bigint | 20 | 是 | 否 | 否 | 主键 |
| UserName | Varchar | 20 | 否 | 是 | 否 | 用户名 |
| NickName | Varchar | 60 | 否 | 否 | 否 | 昵称 |
| Passwd | Varchar | 40 | 否 | 否 | 否 | 密码 |
| Email | varchar | 255 | 否 | 否 | 否 | 电子邮件 |
| Question | varchar | 255 | 否 | 否 | 否 | 提示问题 |
| Answer | varchar | 255 | 否 | 否 | 否 | 提示答案 |
| RegTime | datetime |  | 否 | 否 | 否 | 注册时间 |
| LoginTime | datetime |  | 否 | 否 | 否 | 最后登录时间 |
| LoginIP | varchar | 20 | 否 | 否 | 否 | 最后登录 IP |
| LoginTimes | int | 11 | 否 | 否 | 否 | 登录次数 |
| StayTime | bigint | 20 | 否 | 否 | 否 | 停留时间 |
| SignDetail | text |  | 否 | 否 | 是 | 用户签名 |
| HavePic | tinyint | 1 | 否 | 否 | 是 | 是否有头像 |
| PicFileName | varchar | 255 | 否 | 否 | 是 | 头像地址 |
| TimeZone | varchar | 20 | 否 | 否 | 是 | 地区 |

（2）论坛文章表

论坛文章表存放的是用户所发表的文章信息，也包括回复信息，如表 2-3 所示。而在数据库的设计中主要问题是鉴别帖子信息和回复信息。

表 2-3 bbs_forum

| 字 段 名 | 数据类型 | 长 度 | 主 键 | 外 键 | 可 空 | 说 明 |
| --- | --- | --- | --- | --- | --- | --- |
| ID | bigint | 20 | 是 | 否 | 否 | 主键 |
| ParentID | bigint | 20 | 否 | 否 | 否 | 父帖 ID |
| MainID | bigint | 20 | 否 | 否 | 否 | 主帖 ID |
| BoardID | bigint | 20 | 否 | 否 | 否 | 版块 ID |
| BoardName | varchar | 60 | 否 | 否 | 否 | 版块名字 |
| ReNum | int | 11 | 否 | 否 | 否 | 回复数量 |
| ClickNum | int | 11 | 否 | 否 | 否 | 点击次数 |
| UserID | bigint | 20 | 否 | 否 | 否 | 发表者 ID |
| UserName | varchar | 20 | 否 | 否 | 否 | 发表者姓名 |
| NickName | varchar | 60 | 否 | 否 | 否 | 发表者昵称 |
| Title | varchar | 150 | 否 | 否 | 否 | 文章标题 |
| Detail | text | | 否 | 否 | 否 | 文章内容 |
| Sign | text | | 否 | 否 | 否 | 文章说明 |
| ArtSize | int | 11 | 否 | 否 | 否 | 文章字节数 |
| postTime | bigint | 20 | 否 | 否 | 否 | 发表时间 |
| LastTime | bigint | 20 | 否 | 否 | 否 | 最后回复时间 |
| IPAddress | varchar | 20 | 否 | 否 | 否 | 发表者 IP |
| IsNew | tinyint | 1 | 否 | 否 | 否 | 是否是新 |
| IsHidden | tinyint | 1 | 否 | 否 | 否 | 是否隐藏 |
| IsTop | tinyint | 1 | 否 | 否 | 否 | 是否置顶 |
| IsLock | tinyint | 1 | 否 | 否 | 否 | 是否被锁定 |

（3）论坛版块表

论坛版块表中存放的是论坛中版块信息。而论坛也划分为父论坛和子论坛，区分它们也靠表中的某个字段来实现的。论坛版块表的具体设计如表 2-4 所示。

表 2-4 bbs_board

| 字 段 名 | 数据类型 | 长 度 | 主 键 | 外 键 | 可 空 | 说 明 |
| --- | --- | --- | --- | --- | --- | --- |
| ID | bigint | 20 | 是 | 否 | 否 | 主键 |
| ParentID | bigint | 20 | 否 | 否 | 否 | 父版块 ID |
| ChildIDs | varchar | 255 | 否 | 否 | 否 | 子版块 ID |
| BoardName | varchar | 255 | 否 | 否 | 否 | 版块名字 |
| Explains | text | | 否 | 否 | 否 | 版块描述 |
| BoardPic | varchar | 200 | 否 | 否 | 否 | 图片地址 |
| Orders | int | 11 | 否 | 否 | 否 | 显示顺序 |
| IsHidden | tinyint | 1 | 否 | 否 | 否 | 是否隐藏 |
| PostNum | int | 11 | 否 | 否 | 否 | 文章数量 |

## 2.4 小 结

搭建 Java Web 应用运行环境是学习 JSP 技术的基础。本章介绍了 JDK 和 Tomcat 的安装配置，以及 Eclipse 开发平台的安装与服务器配置。通过本章学习，读者可以使用 Eclipse 开发部署 Java Web 应用项目。

# 第 3 章 JSP 基础

本章需要了解 JSP 页面的基本结构、指令和动作标签、JSP 的运行原理，及怎样设计 JSP 页面；熟悉 JSP 基本语法，并能熟练操作；掌握 JSP 基本语法，包括标准语法如注释、声明、表达式和程序段等；掌握 JSP 的指令标签和动作标签及 JSP 主要元素的使用方法。

## 3.1 JSP 简介

JSP 是基于 Java Servlet 以及整个 Java 体系的 Web 开发技术，利用这一技术可以建立先进、安全、快速和跨平台的动态网站。JSP 技术是用 Java 语言作为脚本语言的，JSP 网页为整个服务器端的 Java 库单元提供了一个接口来服务于 HTTP 的应用程序。

JSP 是一种很容易学习和使用的、在服务器端编译执行的 Web 设计语言。在传统的 HTML 网页文件中加入 Java 程序片段和 JSP 标记，就构成了 JSP 网页。在 JSP 环境下，HTML 代码主要负责描述信息的显示样式，而程序代码则用来描述处理逻辑。普通的 HTML 页面只依赖于 Web 服务器，而 JSP 页面需要附加的语言引擎分析和执行程序代码。程序代码的执行结果被重新嵌入 HTML 中，然后一起发送给浏览器。

JSP 页面的用途就是获取一个请求，再为它创建一个响应信息。Web 服务器接收到访问 JSP 网页的请求时，首先执行其中的程序片段，然后将执行结果以 HTML 格式返回给客户。程序片段可以操作数据库、重新定向网页以及发送 E-mail 等，这就是建立动态网站所需要的功能。所有程序操作都在服务器端执行，网络上传送给客户端的仅是得到的结果，对客户浏览器的要求很低。

### 3.1.1 JSP 页面的基本结构

JSP 页面由以下元素构成：

· 静态内容：这些静态内容是 JSP 页面中的静态文本，它基本上是 HTML 文本，与 Java 和 JSP 语法无关。

· 指令：JSP 指令语法为<%@ 属性=值 %>。

· 表达式：JSP 表达式语法为<%=表达式%>。

· 程序片（Scriptlet）：Scriptlet 是嵌在页面里的一段 Java 代码，语法为<% Java 代码 %>。

· 声明：JSP 声明用于定义 JSP 页面中的变量和方法，语法为<%! 变量和方法%>。

· 动作：JSP 动作允许在页面间转移控制权，语法为<jsp:动作名 属性=值 />或者为<jsp:动作名 属性=值>Body </jsp:动作名>。

· 注释：注释有两种，语法为<!--这种注释客户端可以看到-->和<%--这种注释客户端不能查看到--%>。

JSP 程序例子如下：

```
jiegou.jsp
<%@ page contentType="text/html; charset=gb2312" language="java"%>
```

```
<head>
<meta http-equiv="Content-Type" content="text/html; charset=gb2312" />
<title>JSP 页面结构</title>
</head>
<%! String str="Hello,Everyone!" %>          //声明
<body>
<center>
    <%= new java.util.Date() %>              //表达式
    <% out.println(str) ; %>                 //代码片段
</center>
</body>
</html>
```

JSP 使得我们能够分离页面的静态 HTML 和动态部分。HTML 可以用任何通常使用的 Web 制作工具编写，编写方式也和原来的一样，动态部分的代码放入特殊标记之内，大部分以<%开始，以%>结束。

除了普通 HTML 代码之外，嵌入 JSP 页面的其他成分主要有如下 3 种：脚本元素（Scripting Element），指令元素（Directive），动作元素（Action）。脚本元素用来嵌入 Java 代码，这些 Java 代码将成为转换得到的 Servlet 的一部分；JSP 指令元素用来从整体上控制 Servlet 的结构；动作元素用来引入现有的组件或者控制 JSP 引擎的行为，如表 3-1 所示。

表 3-1　JSP 元素一览表

| 元素类型 | JSP 元素 | 语　法 | 解　释 |
| --- | --- | --- | --- |
| 脚本元素 | 表达式 | <%= 表达式 %> | 表达式经过运算然后输出到页面 |
|  | 程序片 | <% 代码 %> | 嵌入 servlet service 方法中的代码 |
|  | 声明 | <%! 声明代码%> | 嵌入 servlet 中，定义于 service 方法外 |
|  | 注释 | <%-- 注释 --%> | 在将 JSP 转译成 servlet 时将被忽略 |
| 指令元素 | JSP 页面指令 | <% @ page 属性名="值"%> | 在载入时提供 JSP 引擎使用 |
|  | JSP 包含指令 | <% @ include file="url"%> | 在经过转译成 servlet 之后被包含进来的文件 |
| 动作元素 | jsp:include | <jsp:include page=" 相对路径 " flush= " true " > | 在 JSP 中动态包含一个文件 |
|  | jsp:forward | <jsp:forward page="相对路径"> | 将页面得到的请求转向下一页 |
|  | jsp:param | <jsp:include > | 以 "名字—值" 的形式为其他标签提供附加信息 |
|  | jsp:useBean | <jsp:useBean attr="value" > | 找到并建立 JavaBean 对象 |

为了简化脚本元素，JSP 定义了一组可以直接使用的内置对象，可使用的内置对象有 request、response、out、session、application、config、pageContext、exception 和 page。

### 3.1.2　JSP 的运行原理

JSP 页面的扩展名为.jsp，web 服务器通过此扩展名通知 JSP 引擎处理该页面中的元素，通过 JSP 引擎解释 JSP 页面中的标签，生成所需内容。例如，调用一个 JavaBean 来访问一个使用 JDBC API 的数据库或者包含一个文件，然后 JSP 引擎把返回的结果以 HTML 页面的形式发送到浏览器，实质上是把生成内容的业务逻辑封装在服务器端处理的标签和 beans 中。JSP 页面通常被编译成 Java 平台 servlet 类。因此，JSP 页面的运行需要有能支持 Java 平台 servlet 规范的 Java 虚拟机。

当客户端发送一个 JSP 的访问请求时，Web 服务器会把这个请求转发给 JSP 容器，然后 JSP 容器将决定由哪个 JSP 页面实现类来处理这个请求。然后，JSP 容器调用 JSP 页面实现类的一个方法对这个请求进行处理，并且通过容器和 Web 服务器把响应返回给客户端。通常，这个过程可以简单说成"向 JSP 发送一个请求"。

JSP 运行原理如下：当 Web 服务器上的 JSP 页面第一次被请求执行时，JSP 引擎先将 JSP 页面文件转译成一个 Java 文件，即 Servlet（Java Servlet 是基于服务器端编程的 API，用 Java Servlet 编写的 Java 程序称为 Servlet，Servlet 通过 HTML 与客户交互）。服务器将前面转译成的 Java 文件编译成字节码文件，再执行这个字节码文件来响应客户的请求。当这个 JSP 页面再次被请求时，只要该 JSP 文件没有被改动，JSP 引擎就直接调用已装载的 Servlet，如图 3-1 所示。

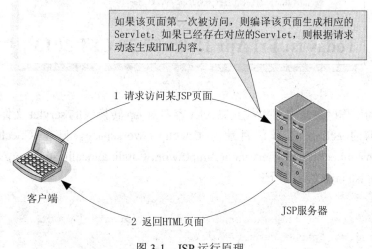

图 3-1  JSP 运行原理

JSP 的运行过程如下：

（1）JSP 引擎先把该 JSP 文件转换成一个 Java 源文件（Servlet），在转换时如果发现 JSP 文件的语法错误，转换过程将中断，并向服务器和客户端输出错误信息。

（2）如果转换成功，JSP 引擎用 Javac 把该 Java 源文件编译成相应的 class 文件。

（3）创建一个该 Servlet 的实例，执行该 Servlet 的 jspInit()方法，jspInit()方法在 Servlet 的生命周期中只被执行一次。

（4）jspService()方法被调用来处理客户端的请求。对每一个请求，JSP 引擎创建一个新的线程来处理该请求。如果有多个客户端同时请求该 JSP 文件，则 JSP 引擎会创建多个线程。每个客户端请求对应一个线程。以多线程方式执行可以大大降低对系统的资源需求，提高系统的并发量及响应时间。

（5）如果该 JSP 文件被修改了，服务器将根据设置决定是否对该文件重新编译，如果需要重新编译，则将编译结果取代内存中的 Servlet，并继续上述处理过程。

（6）在任何时候如果由于系统资源不足的原因，JSP 引擎将以某种不确定的方式将 Servlet 从内存中移去。当这种情况发生时 jspDestory()方法首先被调用，然后 Servlet 实例便被标记并加入"垃圾收集"处理。

（7）可在 jspInit()中进行一些初始化工作，如建立与数据库的连接、建立网络连接、从配置文件中取一些参数等，在 jspDestory()中释放相应的资源。

**例 3-1**：测试第一个 JSP 页面。

```
first_jsp
<%@ page language="java" contentType="text/html; charset=UTF-8"
    pageEncoding="UTF-8"%>
<html>
<body>
<h3>Today is: <%=new java.util.Date() %></h3>
</body>
</html>
```

运行结果如图 3-2 所示。

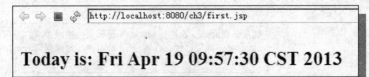

图 3-2　运行效果

在 eclipse 的工作区目录中可以找到 servlet 容器对 jsp 转换成的 servlet 文件 first_jsp.java。

例如：当前的 eclipse 工作区目录为 C:\eclipse\workspace，则在 C:\eclipse\workspace\.metadata\.plugins\org.eclipse.wst.server.core\tmp0\work\Catalina\localhost\ch3\org\apache\jsp 目录下可找到 first_jsp.java 代码如下：

```java
/*
 * Generated by the Jasper component of Apache Tomcat
 * Version: Apache Tomcat/7.0.25
 * Generated at: 2013-04-19 02:10:22 UTC
 * Note: The last modified time of this file was set to
 *       the last modified time of the source file after
 *       generation to assist with modification tracking.
 */
package org.apache.jsp;

import javax.servlet.*;
import javax.servlet.http.*;
import javax.servlet.jsp.*;

public final class first_jsp extends org.apache.jasper.runtime.HttpJspBase
    implements org.apache.jasper.runtime.JspSourceDependent {

  private static final javax.servlet.jsp.JspFactory _jspxFactory =
        javax.servlet.jsp.JspFactory.getDefaultFactory();

  private static java.util.Map<java.lang.String,java.lang.Long> _jspx_dependants;

  private javax.el.ExpressionFactory _el_expressionfactory;
  private org.apache.tomcat.InstanceManager _jsp_instancemanager;
```

```java
  public java.util.Map<java.lang.String,java.lang.Long> getDependants() {
    return _jspx_dependants;
  }

  public void _jspInit() {
    _el_expressionfactory = _jspxFactory.getJspApplicationContext(getSer-
       vletConfig().getServletContext()).getExpressionFactory();
    _jsp_instancemanager = org.apache.jasper.runtime.InstanceManagerFactory.
       getInstanceManager(getServletConfig());
  }

  public void _jspDestroy() {
  }

  public void _jspService(final javax.servlet.http.HttpServletRequest request,
        final javax.servlet.http.HttpServletResponse response)
        throws java.io.IOException, javax.servlet.ServletException {

    final javax.servlet.jsp.PageContext pageContext;
    javax.servlet.http.HttpSession session = null;
    final javax.servlet.ServletContext application;
    final javax.servlet.ServletConfig config;
    javax.servlet.jsp.JspWriter out = null;
    final java.lang.Object page = this;
    javax.servlet.jsp.JspWriter _jspx_out = null;
    javax.servlet.jsp.PageContext _jspx_page_context = null;


    try {
      response.setContentType("text/html; charset=UTF-8");
      pageContext = _jspxFactory.getPageContext(this, request, response,
                 null, true, 8192, true);
      _jspx_page_context = pageContext;
      application = pageContext.getServletContext();
      config = pageContext.getServletConfig();
      session = pageContext.getSession();
      out = pageContext.getOut();
      _jspx_out = out;

      out.write("\r\n");
      out.write("<html>\r\n");
      out.write("<body>\r\n");
      out.write("<h3>Today is: ");
      out.print(new java.util.Date() );
      out.write("</h3>\r\n");
      out.write("</body>\r\n");
      out.write("</html>");
    } catch (java.lang.Throwable t) {
```

```
        if (!(t instanceof javax.servlet.jsp.SkipPageException)){
         out = _jspx_out;
         if (out != null && out.getBufferSize() != 0)
           try { out.clearBuffer(); } catch (java.io.IOException e) {}
         if (_jspx_page_context != null) _jspx_page_context.handlePageException(t);
        }
      } finally {
        _jspxFactory.releasePageContext(_jspx_page_context);
      }
    }
}
```

## 3.2 JSP 语法

### 3.2.1 变量和方法的声明

JSP 声明变量和方法的作用是：在 JSP 程序中声明合法的变量和方法，定义插入 Servlet 类的方法和成员变量。

JSP 声明标签的格式：<%! declaration; [ declaration; ] ... %>

例子：

```
<%! int i = 0; %>
<%! int i, j, k; %>
<%! Circle a = new Circle(2.0); %>
```

描述：

JSP 程序中，在要使用一个变量或引用一个对象的方法和属性前，必须对要使用的变量和对象进行声明，声明后，就可以在后面的程序中使用它们了，还可以声明多个变量或对象实例，但在每个声明后都要用分号结尾。除了可以使用在 JSP 程序中声明变量和对象实例外，还可以使用通过导入的包中声明的变量和对象，这些变量和对象可以直接使用，而不用再在 JSP 程序中声明。

由于声明不会有任何输出，因此它们往往和 JSP 表达式或 scriptlet 结合在一起使用。例如，下面的 JSP 代码片断输出自从服务器启动或 Servlet 类被改动并重新装载以来，当前页面被请求的次数。

```
<%! private int accessCount = 0; %>
```

自从服务器启动以来页面访问次数为：

```
<%= ++accessCount %>
```

注意：

（1）在<%!和%>之间声明变量或方法，其中变量和方法可以是 Java 中的任何变量，其作用范围是整个 JSP 文件。当多个用户访问这个页面时，它们将共享这些变量，任何对变量的修改都会直接影响到其他用户。这些变量只有服务器关闭时才会释放。

例 3-2：利用声明的变量实现访问网页计数功能。

```
exp_2.jsp
<%@ page language="java" contentType="text/html; charset=UTF-8"
    pageEncoding="UTF-8"%>
<head>
<title>jsp 声明</title>
</head>
<body>
    <%!int i=0;%>
    <%i++;%>
    <P>您是第 <%=i%> 个访问本站的客户。
</body>
</html>
```

运行结果如图 3-3、图 3-4 所示。

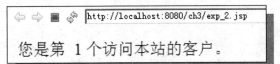

图 3-3  第 1 次运行时该页面显示结果　　　　图 3-4  点击刷新按钮后计数增加显示结果

另外，如果用这种方式来声明方法，则其中方法在整个 JSP 页面有效。但在该方法中声明的变量为局部变量，只在该方法中有效，调用完毕时即释放。而用户往往会调用方法来修改某个变量，所以，在应用这种方法声明变量时，如果试图修改某个数据，尽量加上关键字 synchronized 以实现同步锁定，防止同时修改变量。

利用这种方式声明的对象，也是在 JSP 页面内有效，可以在其范围内使用该对象，具体方法和前面的一样。

（2）在<%和%>之间声明变量和方法。使用此种方式将声明为局部变量和方法，可以简单地理解为非共享变量，即每个客户都有自己的变量，他们的操作不会影响到其他人，程序执行完内存自动释放。大多的 Java 程序片是放在这里的。

（3）尽量不要在 JSP 中过多使用前两种方式来进行动态操作，应该尽量使用 JavaBean 来操作，尽量做到将静态页面和动态页面相分离，这也是为什么要用 JSP 替代 Servlet 的原因。

### 3.2.2  JSP 表达式

JSP 表达式标签的作用：把 Java 数据直接插入到输出。

JSP 表达式标签的格式：<%= Java Expression %>

描述：

计算 Java 表达式得到的结果被转换成字符型数据，正因为这一点，表达式的结果就可以作为 HTML 的内容显示在浏览器窗口中了。计算在运行时（页面被请求时）进行，因此可以访问和请求有关的全部信息。例如，下面的代码显示页面被请求的日期/时间：

```
Current time: <%= new java.util.Date() %>
```

为简化这些表达式，JSP 预定义了一组可以直接使用的内置对象。后面我们将详细介绍这些隐含声明的对象，但对于 JSP 表达式来说，最重要的几个对象及其类型如下：

```
request: HttpServletRequest
```

```
response: HttpServletResponse
session: 和 request 关联的 HttpSession
```

out：PrintWriter（带缓冲的版本，JspWriter），用来把输出发送到客户端
例如：

```
Your hostname: <%= request.getRemoteHost() %>
```

**例 3-3**：使用表达式显示系统的当前时间。

```
exp_3.jsp
<%@ page language="java" contentType="text/html; charset=UTF-8"
    pageEncoding="UTF-8"%>
<html>
<head>
<title>JSP 表达式</title>
</head>
<body>
Hello! The time is now :<%= new java.util.Date() %>
</body>
</html>
```

注意：每次在浏览器中重载网页的时候，它就显示当前时间。标签<%=和%>的作用是标识 Java 表达式的范围，这个表达式将在运行的时候被计算。正因为这样，使用 JSP 产生动态 HTML 网页来响应用户的动作才变为可能。

注意：

（1）不能在一个表达式的末尾添加分号，这点和前面提到的 JSP 声明不同，在声明中，每个声明的结尾都要使用分号结尾。

（2）可以使用任何 Java 语言允许使用的任何表达式。

（3）表达式可以是一个复杂的表达式，这样的表达式是按照从左向右的方式来处理的。

### 3.2.3　Java 程序片（scriptlet）

Java 程序片标签的作用：可以插入 Java 程序代码。如果要完成的任务比插入简单的表达式更加复杂，可以使用 JSP scriptlet。JSP scriptlet 允许把任意 Java 代码插入 Servlet。

Java 程序片标签的格式：　<% Java code; %>

描述：

一个 JSP 页面可以有许多程序片，这些程序片将被 JSP 引擎按顺序执行。在一个程序片中声明的变量称作 JSP 页面的局部变量，它们在 JSP 页面内的所有程序片部分以及表达式部分内有效。这是因为 JSP 引擎将 JSP 页面转译成 Java 文件时，将各个程序片中的变量作为类中某个方法的变量，即局部变量。利用程序片的这个性质，可以将一个程序片分割成几个更小的程序片，然后在这些小的程序片之间再插入 JSP 页面的其他标记元素。当程序片被调用执行时，程序片中的变量被分配内存空间，当所有的程序片调用完毕，这些变量即可释放所占的内存。当多个客户请求同一个 JSP 页面时，JSP 引擎为每个客户启动一个线程，一个客户的局部变量和另一个客户的局部变量会被分配不同的内存空间。因此，一个客户对 JSP 页面局部变量操作的结果，不会影响其他客户的这个局部变量。

**例 3-4**：循环显示表格。

```jsp
exp_4.jsp
<%@ page language="java" contentType="text/html; charset=UTF-8"
    pageEncoding="UTF-8"%>
<html>
<body>
<table width="250 px" border="1">
<% for(int i=0;i<6;i++){ %>
<tr> <td>这是第 <%= i %> 列表格显示</td> </tr>
<% } %>
</table>
</body>
</html>
```

运行结果如图 3-5 所示。

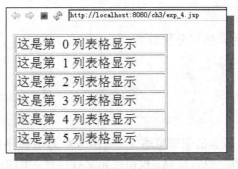

图 3-5　运行效果图

### 3.2.4　JSP 注释

适当地在程序中加入注释可以增强程序的可读性，便于维护人员的维护。即便对于程序编写人员本人，注释对调试程序和编写程序亦可起到很好的帮助作用。程序员在程序中书写注释是一种良好的习惯，在平时学习和实践时要注意培养这种良好的素养。

**1. HTML 注释**

在符号<!--与-->之间的是 HTML 注释，Web 服务器不会作为程序代码解释 HTML 注释，而直接把它交给客户端的浏览器，因此这种注释在浏览器中直接查看 JSP 文件的源文件时可以看到。

格式如下：

```
<!--    注释内容    -- >
```

例如：JSP 使用 HTML 注释，在客户端显示一个注释。

```jsp
<%@page language="java" %>
<html>
    <head>
        <title>HTML 注释测试</title>
    </head>
```

```
    <body>
        <h3>HTML注释测试</h3>
        <!-- a HTML comment example. -->
    </body>
</html>
```

例如：使用表达式的 HTML 注释。

```
<!-- This page was loaded on <%= (new java.util.Date()).toLocaleString() %> -->
```

在客户端的 HTML 源代码中显示如下：

```
<!-- This page was loaded on January 1, 2008 -->
```

这种注释和 HTML 中的很像，也就是它可以在"查看源代码"中看到。唯一有些不同的就是，在这个注释中可以用表达式。这个表达式是不定的，因页面不同而不同，能够使用各种表达式，只要是合法的就行。

2. JSP 注释

JSP 注释的方法有 4 种，格式分别如下：

```
<%-- 注释内容 --%>
// 注释内容
/* 注释内容 */
/** 注释内容 */
```

第一种注释会被 Web 服务器引擎忽略，一般用来对 Java 程序片作出说明；第二种注释是单行注释；第三种方式可以是单行注释，也可以是多行注释；第四种方式是 Java 所特有的 doc 注释。

**例 3-5**：JSP 注释，写在 JSP 程序中，但客户不可见。

```
exp_5.jsp
<%@ page language="java" contentType="text/html; charset=UTF-8"
    pageEncoding="UTF-8"%>
<html>
<head><title>A Comment Test</title></head>
<body>
<h2>A Test of Comments</h2>
<%-- This comment will not be visible in the page source --%>
</body>
</html>
```

运行结果如图 3-6 所示。

描述：

用 JSP 注释标记的字符会在 JSP 编译时被忽略掉。这个注释在希望隐藏或注释 JSP 程序时是很有用的。JSP 编译器不会对<%--标签和--%>标签之间的语句进行编译，它不会显示在客户的浏览器中，在源代码中也不会看到。

图 3-6 运行效果图

## 3.3 JSP 指令标签

### 3.3.1 page 指令

page 指令作用是：定义当前 JSP 文件及其之后的一个或多个 JSP 文件的全局属性（这些属性是区分大小写的）。

page 指令标签格式如下：

```
<% @ page
  [ language = "java" ]
  [ extends = "package.class" ]
  [ import = " { package.class|package.* }, ..." ]
  [ session = "true|false" ]
  [ buffer = "none|8kb|sizekb" ]
  [ autoFlush = "true|false" ]
  [ isThreadSafe = "true|false" ]
  [ info="text" ]
  [ errorPage = "relativeURL" ]
  [ isErrorPage = "true|false" ]
  [contentType = "mimeType[;charset=characterSet]"|"text/html;charset=
              ISO-8859-1"]
%>
```

例如：

```
<%@ page info = "this is a jsp page"%>
<%@ page language = "java"  import = "java.sql.* " %>
<%@ page import = "java.sql.*, java.util.List " %>
<%@ page contentType="text/html; charset=gb2312" language="java"
                import="java.sql.*" errorPage="" %>
```

描述：

- language="java"

主要指定 JSP Container 要用什么语言来编译 JSP 网页。默认值为 Java。

- extends="package.class"

定义此 JSP 编译后产生的 Servlet 是继承哪个父类。

- import="package.class"或者 import="package.class1，...，package.classN

用于指定该 JSP 网页使用到的 Java API（即用于指定 JSP 编译时需要导入的 Java 包，这些包将作用于程序段、表达式以及声明）。

例如：<%@ page import="java.util.*" %>将指定导入 Java 包 java.util.*。

就像 Java 源文件一样，嵌入 JSP 页面中的 Java 代码必须导入代码所用到的类包。如果有多个包则用逗号隔开；例如：import="java.io.*，java.util.*"。

注意：下面的包在 JSP 编译时是默认导入的，不需要再指明了。

```
java.lang.*
javax.servlet.*
```

```
javax.servlet.JSP.*
javax.servlet.http.*
```
• session="true|false"

定义此 JSP 网页是否可以使用 session 对象。如果值为 true，那么该页面加入一个会话；false 表示不加入会话，也不能访问任何会话信息。默认值为 true。

• buffer="none|8kb|sizekb"

定义执行后的 JSP 对客户端浏览器的输出流的缓冲区大小。默认值和服务器有关，但至少应该是 8KB。默认值是 8KB。

• autoFlush="true|false"

定义输出流的缓冲区是否要自动清除。如果 buffer 设置为 none，那么必须把 autoFlush 设置为 false；如果 buffer 没有设置为 none，那么最好把 autoFlush 设置为 true，因为缓冲区满了会产生异常。默认值为 true。

• isThreadSafe="true|false"

用来设置 JSP 文件是否能多线程使用，表示容器是否向该页面并发地传递请求。有效值为 true 和 false。默认值为 true。如果为 true，容器可以允许多个客户端并发访问 JSP。如果为 false，那么容器只能按照接收的次序一次传递一个请求，而页面编写者还必须保证访问共享资源时是完全异步的。

• info="text"

任意字符串。这个属性可以是任何值。JSP 可以把它当做一个管理工具，提供页面的内容、目的、名称等信息

• errorPage="relative URL"

如果页面出现错误，那么应当向客户端发送一个用来显示错误的 Web 页面的 URL。默认的 URL 是依赖于实现的。如果你没有给出一个 URL，那么容器将使用它自己默认的 URL。

• isErrorPage="true|false"

表示当前页面是否是一个错误页。默认是 false。

例 3-6：JSP 中的错误页面。

```
exp_6_1.jsp
<%@ page language="java" contentType="text/html; charset=UTF-8"
    pageEncoding="UTF-8" errorPage="exp_6_2.jsp" %>
<html>
<body>
<你将看不到这些字，因为下面抛出了一个异常信息。>
<%
    int x = 2, y = 0;
    int z = x/y;
%>
</body>
</html>


exp_6_2.jsp
<%@ page language="java" contentType="text/html; charset=UTF-8"
```

```
        pageEncoding="UTF-8" isErrorPage="true" %>
<html>
<body>
<h1>错误</h1>
当处理您的请求时发生了一个错误。
<p>
错误信息为: <%= exception.getMessage() %>。
</body>
</html>
```

运行结果如图 3-7 所示。

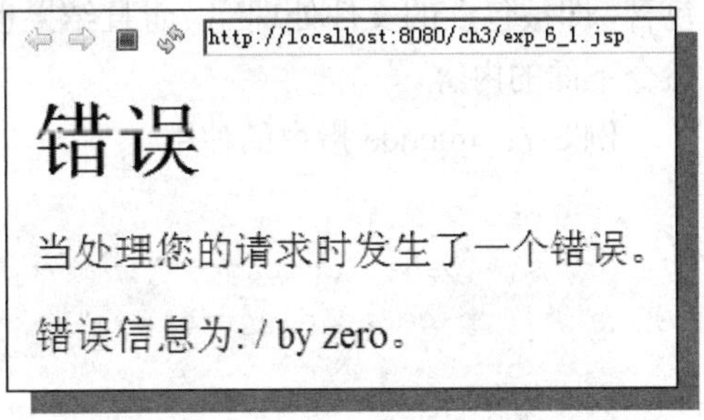

• contentType="mimeType[;charset=characterSet]"|"text/html;charset=ISO-8859-1"

定义页面内容的类型。内容的类型可以是一个简单的类型规范，或者一个类型规范和一个字符集（charset）。JSP 风格的标签的默认值是 text/html。当包含字符集时，该属性的语法是 contentType="text/html;charset=char set identifier"。在属性值中分号之后可以有空格。因为字符集说明了字符如何编码，所以页面可以用不同的脚本来支持各种语言。关于字符集的一些信息可以参考以下网页 http://www.w3.org/TR/REC‾html40/charset.html。

图 3-7  运行 exp_6_1.jsp 页面效果图

较常见的错误如下：

```
<%@ page language="java" contentType="text/html"; charset ="GB2312"%>
```

应改为：

```
<%@ page language="java" contentType="text/html; charset= GB2312" %>
```

注意：

（1）无论把<%@ page ……%>指令放在 JSP 文件的哪个地方，它的作用范围都是整个 JSP 页面。不过，为了 JSP 程序的可读性以及好的编程习惯，最好还是把它放在 JSP 文件的顶部。

（2）可以在一个页面中用多个<%@ page ……%>指令，但是其中的属性只能用一次。不过也有个例外，那就是 import 属性，因为 import 属性和 Java 中的 import 语句差不多，所以此属性能多用几次。

（3）<%@ page …%>指令作用于整个 JSP 页面，同样包括静态的包含文件。但是<%@ page …%>指令不能作用于动态的包含文件，比如<jsp:include>。

### 3.3.2 include 指令

include 指令标签作用是：包含一个文本或代码的文件。
include 指令标签格式如下：

```
<%@ include file="relative URL"%>
```

描述：

include 指令只有一个 file 属性，用于在代码当前位置指定包含的文件名。被包含的文件可以是任何 HTML 或者 JSP 文件，甚至可以是代码片段。file 属性的值是一个 URL 的一部分，没有协议、端口号或域名，其路径是相对于 JSP 文件而言，它指向 Web 应用中的某个文件。例如：

```
"error.jsp"
"/templates/onlinestore.html"
"/beans/calendar.jsp"
```

如果被包含的文件是 JSP 文件，则 JSP 引擎编译完该 JSP 文件后，执行的结果将插入主 JSP 文件中 include 指令所在的位置。如果被包含的文件是 HTML 文件或文本文件，则 JSP 引擎不对其进行编译，直接将其内容插入主 JSP 文件中 Include 指令所在的位置。该包含是静态包含，即被包含的文件处理完，而且结果也插入了主 JSP 文件，主 JSP 文件将继续执行 include 指令下面的内容。

**例 3-7**：inlcude 指令的使用。

```
exp_7_1.jsp
<%@ page language="java" contentType="text/html; charset=UTF-8"
    pageEncoding="UTF-8"%>
<html>
<body>
exp_7_2.jsp中的随机显示的数为：<br/>
<%@ include file="exp_7_2.jsp" %>
</body>
</html>
exp_7_2.jsp
<%@ page language="java" contentType="text/html; charset=UTF-8"
    pageEncoding="UTF-8" %>
<%=java.lang.Math.random()*10000%>
```

在页面中显示如图 3-8 所示。

许多网站的每个页面都有一个小小的导航条。由于 HTML 框架存在不少问题，导航条往往用页面顶端或左边的一个表格制作，同一份 HTML 代码重复出现在整个网站的每个页面上。include 指令是实现该功能的非常理想的方法。使用 include 指

图 3-8 运行 exp_7_1.jsp 页面效果图

令，开发者不必再把导航 HTML 代码复制到每个文件中，从而可以更轻松地完成维护工作。

由于 include 指令是在 JSP 转换成 Servlet 的时候引入文件，因此如果导航条改变了，所有使用该导航条的 JSP 页面都必须重新转换成 Servlet。如果导航条改动不频繁，而且希望包含操作具有尽可能好的效率，使用 include 指令是最好的选择。然而，如果导航条改动非常频繁，则可以使用 jsp:include 动作。jsp:include 动作在出现对 JSP 页面请求的时候才会引用指定的文件，请参见本文后面的具体说明。

注意：

（1）被包含的文件中不能含有<html>、</html>、<body>或</body>标签。因为被包含的文件的全部内容将被插入 JSP 文件中 include 指令所在的地方，这些标签将会同 JSP 文件中已有的同样的标签发生冲突。

（2）假如被包含的文件发生变化，主 JSP 页面将被重新编译。

## 3.4 JSP 动作标签

JSP 动作利用 XML 语法格式的标记来控制 Servlet 引擎的行为。利用 JSP 动作可以动态地插入文件，重用 JavaBean 组件，把用户重定向到另外的页面，为 Java 插件生成 HTML 代码。JSP 动作如下。

jsp:include：在页面被请求的时候引入一个文件。
jsp:forward：把请求转到一个新的页面。
jsp:useBean：寻找或者实例化一个 JavaBean。
jsp:setProperty：设置 JavaBean 的属性。
jsp:getProperty：输出某个 JavaBean 的属性。

### 3.4.1 jsp:include 动作标签

jsp:include 动作标签的作用是：把指定静态或动态文件插入正在生成的页面。
jsp:include 动作标签格式如下：

```
<jsp:include page="relative URL" flush="true" />
```

例子：

```
<jsp:include page="hello.html" />
<jsp:include page="/index.html" />
```

描述：
<jsp:include>标签允许包含一个静态文件或动态文件。一个静态文件被执行后，它的内容插入主 JSP 页面中。一个动态文件对请求作出响应，而且将执行结果插入 JSP 页面中。

<jsp:include>标签能处理两种文件类型，当你不知道这个文件是静态或动态的文件时，使用该标签是非常方便的。

当 include 动作执行完毕后，JSP 引擎将接着执行 JSP 文件剩下的文件代码。

```
•page="{ relative URL | <%= expression %>}"
```

该属性指出被包含文件相关 URL，该 URL 不能包含协议名、端口号或域名。该 URL 是绝对或相对于当前 JSP 文件来说的。如果它是绝对地址（以"/"开始），该路径由 Web 或应用服务器决定。

```
•flush="true"
```

在 JSP 文件中，常设置 flush="true",因为它不是一个默认值。

include 指令是在 JSP 文件被转换成 Servlet 的时候引入文件，而<jsp:include>动作不同，它插入文件的时间是在页面被请求的时候。

### 3.4.2 jsp:forward 动作标签

jsp:forward 动作标签的作用是：重定向一个 HTML 文件、JSP 文件，或者一个程序段。
jsp:forward 动作标签格式如下：

```
<jsp:forward page={"relativeURL" | "<%= expression %>"} />
```

或

```
<jsp:forward page={"relativeURL" | "<%= expression %>"} >
    <jsp:param name="parameterName"
        value="{parameterValue | <%= expression %>}" />
</jsp:forward>
```

例如：

```
<jsp:forward page="/servlet/login" />
<jsp:forward page="/servlet/login">
    <jsp:param name="username" value="zhangsan" />
</jsp:forward>
```

描述：

<jsp:forward>标签从一个 JSP 文件向另一个文件传递一个包含用户请求的 request 对象，能够向目标文件传送参数和值，在这个例子中传递的参数名为 username,值为 zhangsan，如果使用了<jsp:param>标签的话，目标文件必须是一个动态的文件，能够处理参数。

如果使用了非缓冲输出的话，那么使用<jsp:forward>时就要小心。如果在使用<jsp:forward>之前，jsp 文件已经有了数据，那么文件执行就会出错。

属性：

page="{relativeURL | <%= expression %>}"

这里是一个表达式或一个字符串，用于说明将要定向的文件或 URL。这个文件可以是 JSP 程序段，或者其他能够处理 request 对象的文件（如 asp、cgi、php）。

```
<jsp:param name="parameterName" value="{parameterValue | <%= expression %>}" />
```

向一个动态文件发送一个或多个参数，这个文件一定是动态文件。

如果想传递多个参数，可以在一个 JSP 文件中使用多个<jsp:param>。name 指定参数名，value 指定参数值。

### 3.4.3　param 动作标签

param 标签以"名字—值"对应的形式为其他标签提供附加信息，与 jsp:include、jsp:forword 标签一起使用。

param 动作标签的格式如下：

```
<jsp:param name="名字" value="指定给 param 的值">
```

**1. <jsp:param>与<jsp:include>配合使用**

例 3-8：include 动作标签中嵌入 param 动作标签。

```
exp_8_1.jsp
<%@ page language="java" contentType="text/html; charset=UTF-8"
    pageEncoding="UTF-8" %>
<html>
<body>
<% double i=Math.random();%>
<jsp:include page="exp_8_2.jsp">
```

```
    <jsp:param value="<%=i %>" name="number"/>
</jsp:include>
</body>
</html>

exp_8_2.jsp
<%@ page language="java" contentType="text/html; charset=UTF-8"
    pageEncoding="UTF-8" %>
<html>
<body>
<%
//获得 exp_8_1.jsp 传来的值
String str=request.getParameter("number");
double n=Double.parseDouble(str);
%>
从 include 动作标签获取的值为:<br>
<%= n %>
</body>
</html>
```

运行结果如图 3-9 所示。

图 3-9　运行 exp_8_1.jsp 页面效果图

2. <jsp:param>与<jsp:forward>配合使用

**例 3-9**：forward 动作标签中嵌入 param 动作标签模拟用户登录。

```
exp_9_login.jsp
<%@ page language="java" contentType="text/html; charset=UTF-8"
    pageEncoding="UTF-8" %>
<html>
<body>
<form action="exp_9_check.jsp" method="get">
用户: <input type="text" name="username" value=<%=request.getParameter
            ("user") %>><br/>
口令: <input type="password" name="password"><br/>
<input type="submit" value="login">
</form>
</body>
</html>


exp_9_check.jsp
<%@ page language="java" contentType="text/html; charset=UTF-8"
```

```jsp
        pageEncoding="UTF-8" %>
<html>
<body>
<%
String name=request.getParameter("username");
String password=request.getParameter("password");
if(name.equals("admin")&&password.equals("admin")){
%>
<jsp:forward page="exp_9_success.jsp">
    <jsp:param name="user" value="<%=name %>"/>
</jsp:forward>
<% } else { %>
<jsp:forward page="exp_9_login.jsp">
    <jsp:param name="user" value="<%=name %>"/>
</jsp:forward>
<% } %>
</body>
</html>


exp_9_success.jsp
<%@ page language="java" contentType="text/html; charset=UTF-8"
    pageEncoding="UTF-8" %>
<html>
<body>
<%=request.getParameter("user")%>登录成功!
</body>
</html>
```

运行结果如图 3-10、图 3-11 所示。

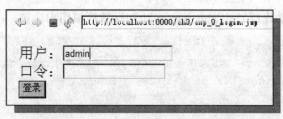

图 3-10　运行 exp_9_login.jsp 页面效果图

图 3-11　登录成功效果图

### 3.4.4　useBean 动作标签

jsp:useBean 动作标签,用来装载一个将在 JSP 页面中使用的 JavaBean。这个功能非常有用,因为它既可以发挥 Java 组件重用的优势,同时也避免了损失 JSP 区别于 Servlet 的方便性。

虽然可以把大段的代码放在脚本片段内，但是绝大多数的 Java 代码属于可重复使用的名为 JavaBean 的组件。JavaBean 能提供常用功能并且可以重复使用。

JavaBean 的值是通过一些属性获得的，可通过这些属性访问 JavaBean 设置。以一个人来类比，一个人就是一个 JavaBean，他的名字、身份证号码以及住址就是其属性。在 JSP 网站，基本上通过应用 Connect the beans 来使网站动态化。

jsp:useBean 动作标签的格式为：

```
<jsp:useBean id="name" class="package.class" scope="page|session|application"/>
```

描述：

这行代码的含义是：创建一个由 class 属性指定的类的实例，然后把它绑定到其名字由 id 属性给出的变量上。jsp:useBean 动作只有在不存在同样 id 和 scope 的 Bean 时才创建新的对象实例，同时，获得现有 Bean 的引用就变得很有必要。

<jsp:useBean>标记要求用 id 属性来识别 Bean。这里提供一个名字来区别 JSP 页面其余部分的 Bean。除了 id 属性，还须告诉网页从何处查找 Bean，或者它的 Java 类别名是什么。这种类别属性提供确认 Bean 的功能，其他一些方法也可以做到这一点。最后一个必需的元素是 scope 属性。有了 scope 属性的帮助，就能告诉 Bean 为单一页面（默认情况）[scope="page"]、为[scope="request"]请求、为会话[scope="session"]，或者为整个应用程序[scope="application"]保持留信息。

一旦声明了一个 JavaBean，就可以访问它的属性来定制它。要获得属性值，可用<jsp:getProperty>标记。有了这个标记，就能指定将要用到的 Bean 名称（从 useBean 的 id 字段得到）以及想得到其值的属性。实际的值被放在输出中：

```
<jsp:getProperty id="localName" property="name" />
```

要改变 JavaBean 属性，必须使用<jsp:setProperty>标记。对这个标记，需要再次识别 Bean 和属性，以修改并额外提供新值。如果命名正确，这些值可以从一个已提交的表中直接获得：参数获得：<jsp:setProperty id="localName" property="*" />；可以从一个参数获得，但必须直接命名属性和参数：<jsp:setProperty id="localName" property="serialNumber" value="string" />；或者直接用一个名字和值来设置：<jsp:setProperty id="localName" property= "serialNumber" value= <%= expression %> />。

例 3-10：装载一个 Bean，然后设置并读取它的 message 属性。

```
 exp_10.jsp
<%@ page language="java" contentType="text/html; charset=UTF-8"
    pageEncoding="UTF-8" %>
<html>
<body>
    <jsp:useBean id="b" class="bean.SimpleBean"/>
    <jsp:setProperty name="b"  property="message"
        value="重新为 message 属性赋值" />
<jsp:getProperty name="b" property="message" />
</body>
</html>
```

exp_10.jsp 页面用到了一个 SimpleBean。

SimpleBean 的代码如下：

```
package bean;
public class SimpleBean {
    private String message="没有内容";
    public String getMessage() {
    return message;
    }
    public void setMessage(String message) {
    this.message = message;
    }
}
```

运行结果如图 3-12 所示。

注意：

为了 Web 服务器能找到 JavaBean，需要将其类别文件放在一个特殊位置。包含 Bean 的类文件应该放到服务器 Java 类的目录下，而不是保留给修改后能够自动装载的类的目录。例如，对于 Java Web Server 来说，Bean 和所有 Bean 用到的类都应该放入 classes 目录，或者封装进 jar 文件后放入 lib 目录，但不应该放到 servlets 下，如图 3-13 所示。

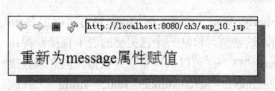

图 3-12  运行 exp_10.jsp 效果图

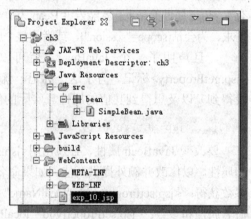

图 3-13  Eclipse 中本例的工程目录图

## 3.5 小  结

本章首先对 JSP 进行了简要介绍，讲解了 JSP 的基本结构以及 JSP 的运行原理；接下来讲解了 JSP 的语法，其中包括 JSP 中变量和方法的声明方法、JSP 表达式、JSP 程序片的使用以及 JSP 的注释方法；最后讲解了 JSP 中的指令标签和动作标签。

## 3.6 习  题

1. 什么是 JSTL，它有什么作用？
2. JSTL 中包含有哪些主要的标签库？
3. JSP 的脚本元素有哪几类？
4. JSP 的指令有什么作用？有哪几种指令？
5. JSP 有哪些标准的动作？

# 第 4 章  JSP 内建对象

JSP 为简化页面的开发提供了一些内建对象，这些内建对象不需要实例化，它们由容器实现和管理。本章主要对 JSP 的 5 种内置对象进行了讲解并给出了具体实例。对这 5 种常用的内置对象，读者需理解每个内置对象提供的功能和使用方法，并会运用其常用的方法。

## 4.1  out 对象

out 对象主要用来向客户端输出各种格式的数据，并且管理应用服务器上的输出缓冲区，out 对象的基类是 javax.servlet.jsp.JspWriter 类。

下面给出了一些 out 对象的常用方法：

out.print/println(boolean|char|double|float|int|long|Object|String|)：用于输出各种类型的数据。out.print()方法与 out.println()方法的区别是，out.print()方法在输出完毕后，并不结束该行，而 out.println()方法在输出完毕后，会结束当前行；下一个输出语句将在下一行开始输出。

out.newLine()：输出一个换行字符。

out.flush()：输出缓冲区里的数据。out.flush()方法也会清除缓冲区中的数据，但是此方法会先将之前缓冲区中的数据输出至客户端，然后再清除缓冲区中的数据。

out.clearBuffer()：清除缓冲区里的是数据，并把数据输出到客户端。

out.clear()：清除缓冲区里的是数据，但不会把数据输出到客户端。

out.getBufferSize()：获得缓冲区的大小。

out.getRemaining()：获取缓冲区中没有被占用的空间的大小。

out.close()：关闭输出流。

另外，out 对象还可以输出一个表格等内容的信息。

下面以几个例子来说明 out 对象的使用方式。

例 4-1：使用 out 对象输出文字信息。

```
out-1.jsp
<%@ page contentType="text/html;charset=GB2312" %>
<%@ page import="java.util.*"%>
<html>
<head>
<title>out 对象的使用例子 1</title>
</head>
<body>
<%  out.println("<h1>这是 h1 字体</h1>");
out.println("<font size=5> 这是 font 标签显示出来的字</font>");
out.println("<center>文字的居中显示</center>");
out.println("<font color=red>文字的红色显示</font>");
int a=10; long b=20; boolean c=true; char d='d'; String e="this is a test!";
out.println(a);
```

```
out.println(b);
out.println(c);
out.println(d);
out.println(e);
%>
</body>
</html>
```

运行结果如图 4-1 所示。

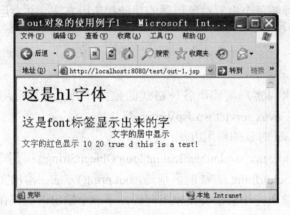

图 4-1  out 对象输出文字信息

**例 4-2**：使用 out 对象输出系统时间信息。

```
out-2.jsp
<%@ page contentType="text/html;charset=GB2312" %>
<%@ page import="java.util.*"%>
<%! public  String  getWeek(int n)
{ String week[]={"星期日","星期一","星期二","星期三","星期四","星期五","星期六"};
     return week[n];
   }
%>
<HTML>
<head>
<title>out 对象的使用例子 1</title>
</head>
<BODY><center><Font size=5 color="#CC00FF">
<% Calendar calendar=Calendar.getInstance(); //创建一个日历对象。
   String yy=String.valueOf(calendar.get(Calendar.YEAR)),
        mm=String.valueOf(calendar.get(Calendar.MONTH)+1),
        dd=String.valueOf(calendar.get(Calendar.DAY_OF_MONTH)),
        ww=getWeek(calendar.get(Calendar.DAY_OF_WEEK)-1);
  int hour=calendar.get(Calendar.HOUR_OF_DAY),
     minute=calendar.get(Calendar.MINUTE),
     second=calendar.get(Calendar.SECOND);
%>
<P>现在时间是:<br><br>
<%=yy%>年
```

```
<%=mm%>月
<%=dd%>日
<%=ww%>
<%=hour%>点
<%=minute%>分
<%=second%>秒
</FONT></center>
</BODY>
</HTML>
```

运行结果如图 4-2 所示。

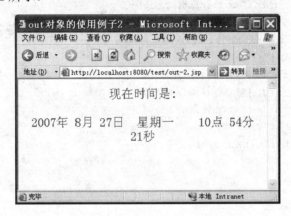

图 4-2  out 对象输出时间信息

例 4-3：使用 out 对象输出表格。

```
out-3.jsp
<%@ page contentType="text/html; charset=gb2312" language="java" import=
                "java.util.*" errorPage="" %>
<html>
<head>
<title>out 对象的使用例子 3</title>
</head>
<body>
<h3>下面是一个课程表：</h3>
<%out.println("<table border>");
  out.println("<tr>");
      out.println("<td>课程表</td>");
      out.println("<td>上午</td>");
      out.println("<td>下午</td>");
  out.println("</tr>");
  out.println("<tr>");
      out.println("<td>周一</td>");
      out.println("<td>语文</td>");
      out.println("<td>数学</td>");
  out.println("</tr>");
  out.println("<tr>");
      out.println("<td>周二</td>");
```

```
        out.println("<td>英语</td>");
        out.println("<td>生物</td>");
    out.println("</tr>");
    out.println("<tr>");
        out.println("<td>周三</td>");
        out.println("<td>化学</td>");
        out.println("<td>物理</td>");
    out.println("</tr>");
    out.println("<tr>");
        out.println("<td>周四</td>");
        out.println("<td>计算机</td>");
        out.println("<td>历史</td>");
    out.println("</tr>");
    out.println("<tr>");
        out.println("<td>周五</td>");
        out.println("<td>音乐</td>");
        out.println("<td>体育</td>");
    out.println("</tr>");
    out.println("</table>");
%>
</body>
</html>
```

运行结果如图 4-3 所示。

图 4-3  out 对象输出表格

## 4.2  request 对象

request 对象用于接收用户请求的所有信息，也就是说，request 对象封装了用户提交的信息，因此，使用该对象调用相应的方法就可以获得用户提交的这些信息。

与 request 相关的是 HttpServletRequest 类。通过 getParameter 方法可以得到 request 参数，获取客户提交信息最常用的方法就是 getParameter（String name）。另外，通过 GET、POST、HEAD 等方法可以得到 request 的类型，通过 cookies、Referer 等可以得到引入的 HTTP 头。在 4.2.3 小节中将给出 request 对象常用方法的实例。

### 4.2.1 获取 HTML 表单提交的数据

使用 HTML 表单提交数据是用户经常用到的提交信息的一种方式，在 HTML 中，用 <form></form> 来添加表单。表单有很多种，使用<input>标签的 type 属性可以进行多种表单元素的添加。

在使用 request 对象的 getParameter（String name）方法时，服务器可以根据 name 属性来获取用户提交的信息。

文本框：<input type="text" name="textfield" />，服务器通过 request 对象的 getParameter 方法的 name 属性得到文本框 value 的初始值或文本框中用户输入的内容。

密码框：<input type="password" name="textfield2" />，服务器通过 name 属性得到密码框中输入的值，密码框仅起到不让别人看到密码的目的，不提供保密措施。

文本域：<textarea name="textarea"></textarea>，服务器通过 name 属性可以得到文本域中用户输入的内容。

单选框：<input type="radio" name="radiobutton" value="radiobutton" />，服务器通过 name 属性得到被选中的单选框的 value 值。

复选框：<input type="checkbox" name="checkbox" value="checkbox" />，服务器通过 name 属性得到被选中的复选框的 value 值。

提交键：<input type="submit" name="Submit" value="提交" />，单击提交键后，服务器可以获得表单提交的各个数据。

重置键：<input type="reset" name="Submit2" value="重置" />，重置键将表单中输入的数据清空，以便重新输入数据。

下拉列表：<select name="select">
    <option>beijing</option>
    <option>henan</option>
    </select>

在下拉列表中，服务器可以通过 name 属性得到被选中的列表的 value 值，即 option 标签中的 value 的值。

**例 4-4**：request-1.jsp 是一个填写用户信息的页面，request-1-answer.jsp 是用 request 对象来获取用户提交的个人信息的页面。

```
request-1.jsp
<%@ page contentType="text/html; charset=gb2312" language="java"
                    import="java.sql.*" errorPage="" %>
<!DOCTYPE html PUBLIC "-//W3C//DTD XHTML 1.0 Transitional//EN"
    "http://www.w3.org/TR/xhtml1/DTD/xhtml1-transitional.dtd">
<html xmlns="http://www.w3.org/1999/xhtml">
<head>
<meta http-equiv="Content-Type" content="text/html; charset=gb2312" />
<title>无标题文档</title>
</head>
<body>
<h2>请填写您的个人信息：</h2>
```

```
<form id="form1" name="form1" method="post" action="request-1-answer.jsp">
  <p>您的姓名:
    <input type="text" name="name" />
  </p>
  <p>您的性别:
    男
    <input name="sex" type="radio" value="male" checked="checked" />
    女
    <input name="sex" type="radio" value="female" />
  </p>
  <p>您所处的城市:
    <select name="city" size="1">
      <option value="beijing">北京</option>
      <option value="zhengzhou">郑州</option>
      <option value="shijiazhuang">石家庄</option>
      <option value="tianjin">天津</option>
      <option value="guangzhou">广州</option>
      <option value="qita">其他</option>
    </select>
  </p>
  <p>您的爱好: 看书
    <input type="checkbox" name="like1" value="read" />
    足球
    <input type="checkbox" name="like2" value="football" />
    音乐
    <input type="checkbox" name="like3" value="music" />
    跳舞
    <input type="checkbox" name="like4" value="dance" />
    旅游
    <input type="checkbox" name="like5" value="travel" />
  </p>
  <p>您的个人简介(顾及到处理汉字的问题,请使用英文填写): </p>
  <p>
    <textarea name="info" cols="40" rows="4" ></textarea>
  </p>
  <p>确认请按提交键:
    <input type="submit" name="submit" value="submit" />
  </p>
</form>
<p> </p>
<p>  </p>
</body>
</html>
request-1-answer.jsp
<%@ page contentType="text/html; charset=gb2312" language="java"
         import="java.sql.*" errorPage="" %>
<!DOCTYPE html PUBLIC "-//W3C//DTD XHTML 1.0 Transitional//EN"
   "http://www.w3.org/TR/xhtml1/DTD/xhtml1-transitional.dtd">
```

```html
<html xmlns="http://www.w3.org/1999/xhtml">
<head>
<meta http-equiv="Content-Type" content="text/html; charset=gb2312" />
<title>无标题文档</title>
</head>
<body>
<h2>您填写的个人信息如下:
</h2>
<% String s1=request.getParameter("name");
   String s2=request.getParameter("sex");
   String s3=request.getParameter("city");
   String s4=request.getParameter("like1");
   String s5=request.getParameter("like2");
   String s6=request.getParameter("like3");
   String s7=request.getParameter("like4");
   String s8=request.getParameter("like5");
   String s9=request.getParameter("info");
   String s10=request.getParameter("submit");
out.print("您的姓名是: ");
out.println(s1);
out.println("<p>");
if (s2==null) {s2="";}
out.print("您的性别是: ");
out.println(s2);
out.println("<p>");
out.print("您所处的城市是: ");
out.println(s3);
out.println("<p>");
out.print("您的爱好是: ");
if (s4==null) {s4="";}
out.println(s4);
if (s5==null) {s5="";}
out.println(s5);
if (s6==null) {s6="";}
out.println(s6);
if (s7==null) {s7="";}
out.println(s7);
if (s8==null) {s8="";}
out.println(s8);
out.println("<p>");
out.print("您的个人简介是: ");
out.println(s9);
out.println("<p>");
out.print("request-1.jsp页面中'提交键'的名称是: ");
out.println(s10);
%>
</body>
</html>
```

运行结果如图 4-4、图 4-5 所示。

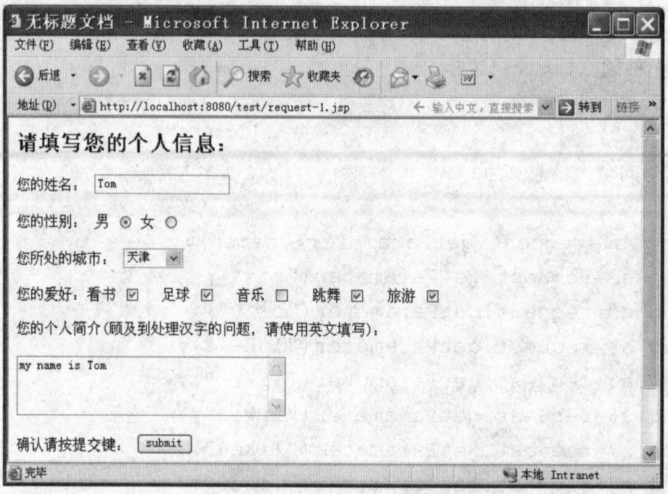

图 4-4 用户填写个人信息页面

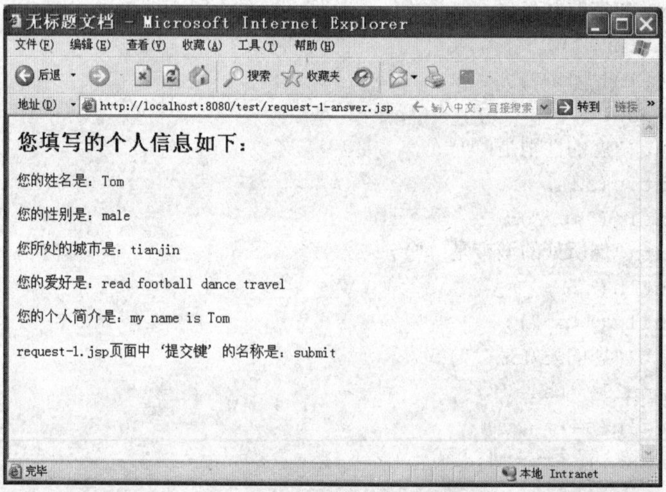

图 4-5 request 对象提取个人信息

**例 4-5**：request-2.jsp 是计算求和的页面，request-2-answer.jsp 是用 request 对象给出结果的页面。

```
request-2.jsp
<HTML>
<%@ page contentType="text/html;charset=GB2312" %>
<BODY>
 <h3>请选择计算和的方式：</h3>
 <FORM action="request-2-answer.jsp" method=post name=form>
      <Select name="calculate" size=1>
         <Option Selected value="1">计算 1 到 n 的连续和
         <Option value="2">计算 1 到 n 的平方和
      </Select>
 <P>请输入 n 的值:
```

```
    <input name="textfield" type="text" size="6">
 <P>
   <INPUT TYPE="submit" value="查看结果" name="submit">
 </FORM>
</BODY>
</HTML>
request-2-answer.jsp
<HTML>
<%@ page contentType="text/html;charset=GB2312" %>
<BODY>
 <% long sum=0;
    String s1=request.getParameter("calculate");
    String s2=request.getParameter("textfield");
    if(s1==null)
     {s1="";}
    if(s2==null)
     {s2="0";}
   if(s1.equals("1"))
     {int n=Integer.parseInt(s2);
      for(int i=1;i<=n;i++)
        {sum=sum+i;
        }
     }
     else if(s1.equals("2"))
      {int n=Integer.parseInt(s2);
       for(int i=1;i<=n;i++)
         {sum=sum+i*i;
         }
      }
  %>
  <h3>您的求和结果是：<%=sum%>
  </h3>
 </BODY>
 </HTML>
```

运行结果如图 4-6、图 4-7 所示。

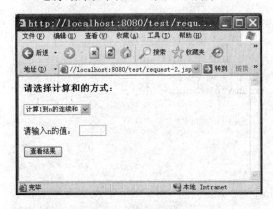

图 4-6  计算求和的页面

图 4-7  request 对象给出结果的页面

## 4.2.2 汉字信息处理

当使用 request 对象来得到汉字或字符信息时，会出现乱码，有两种方法来处理。

方法 1：使用 request 对象的 setCharacterEncoding 方法对字符进行 GBK 编码，那么用 request 对象的 getParameter 方法获得的字符都将是编码后的字符，直接输出即可。

方法 2：首先将获取的字符用 ISO-8859-1 进行编码并将其放到一个字节数组中，然后将这个数组转换为字符串对象即可。

```jsp
String textContent=request.getParameter("text2");
    byte b[]=textContent.getBytes("ISO-8859-1");
textContent=new String(b);
```

**例 4-6**：request-3.jsp 是一个简单的显示页面，request-3-answer1.jsp 是使用方法 1 来处理汉字信息，request-3-answer2.jsp 是使用方法 2 来处理汉字信息。

```jsp
request-3.jsp
<%@ page contentType="text/html;charset=GB2312" %>
<HTML>
<BODY>使用方法1处理汉字信息：
    <FORM action="request-3-answer1.jsp" method=post name=form1>
        <INPUT type="text" name="text1">
        <INPUT TYPE="submit" value="方法1" name="submit1">
    </FORM>
    使用方法2处理汉字信息：
    <FORM action="request-3-answer2.jsp" method=post name=form2>
        <INPUT type="text" name="text2">
        <INPUT TYPE="submit" value="方法2" name="submit2">
    </FORM>
</BODY>
</HTML>

request-3-answer1.jsp
<%@ page contentType="text/html;charset=GB2312" %>
<HTML>
<BODY>
<P>文本框提交的内容是：
    <%request.setCharacterEncoding("GBK");
    String textContent=request.getParameter("text1");
  %><%=textContent%>
<P> 按钮的名字是：
    <%String buttonName=request.getParameter("submit1");
  %><%=buttonName%>
</BODY>
</HTML>

request-3-answer2.jsp
<%@ page contentType="text/html;charset=GB2312" %>
<HTML>
```

```
<BODY>
<P>文本框提交的内容是:
  <%String textContent=request.getParameter("text2");
    byte b[]=textContent.getBytes("ISO-8859-1");
    textContent=new String(b);
  %><%=textContent%>
<P>按钮的名字是:
  <%String buttonName=request.getParameter("submit2");
    byte c[]=buttonName.getBytes("ISO-8859-1");
    buttonName=new String(c);
  %><%=buttonName%>
</BODY>
</IITML>
```

运行结果如图 4-8～图 4-10 所示。

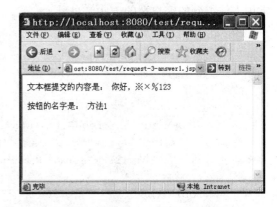

图 4-8　使用两种方法处理汉字信息　　　　　图 4-9　方法 1 处理汉字信息的结果

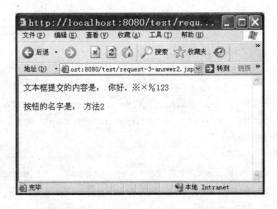

图 4-10　方法 2 处理汉字信息的结果

### 4.2.3　常用方法举例

下面列出了 request 对象常用的一些方法，通过这些方法可以得到客户端的一些信息和客户提交的信息。

getHeader(String name)：获得 HTTP 协议定义的文件头信息，由参数 name 指定头名字。

getHeaderNames()：返回当前请求中所有的报头名称的一个枚举。

getMethod()：获得客户端向服务器端传送数据的方法。
getParameter(String name)：获得客户端传送给服务器端的参数值。
getParameterNames()：获得客户端传送给服务器端的所有参数的名字。
getProtocol()：获取客户端向服务器端传送数据所依据的协议名称。
getRequestURI()：获取发出请求字符串的客户端地址，返回url中主机和端口之后，但在表单数据之前的部分。
getRemoteAddr()：获取客户端的IP地址。
getRemoteHost()：获取客户端的名字。
getServerName()：获取服务器的名字。
getServletPath()：获取客户端所请求的脚本文件的文件路径。
getServerPort()：获取服务器的端口号。

**例 4-7**：request-4.jsp 是一个简单的显示页面，request-4-answer.jsp 是使用 request 对象的常用方法得到的客户端的一些信息显示。

```
request-4.jsp
<%@ page contentType="text/html;charset=GB2312" %>
<%@ page import="java.util.*" %>
<HTML>
<BODY>
  <FORM action="request-4-answer.jsp" method=post name=form>
      <INPUT type="text" name="text">
      <INPUT TYPE="submit" value="查看" name="submit">
  </FORM>
</BODY>
</HTML>

request-4-answer.jsp
<%@ page contentType="text/html;charset=GB2312" %>
<%@ page import="java.util.*" %>
<HTML>
<BODY>
<BR>HTTP 头文件中 accept-encoding 的值(request.getHeader("accept-encoding"))：
  <% String header1=request.getHeader("accept-encoding");
    out.println(header1);
  %>
<BR><BR>HTTP 头文件中 Host 的值(request.getHeader("Host"))：
  <% String header2=request.getHeader("Host");
    out.println(header2);
  %>
<BR><BR>所有报头名称的一个枚举(request.getHeaderNames())：
  <% Enumeration enum_headed=request.getHeaderNames();
    while(enum_headed.hasMoreElements())
        {String s=(String)enum_headed.nextElement();
         out.println(s);
        }
  %>
```

```
<BR><BR>客户提交信息的方式(request.getMethod()):
   <% String method=request.getMethod();
      out.println(method);
   %>
<BR><BR>文本框提交的信息(request.getParameter("text")):
   <% request.setCharacterEncoding("GBK");
      String textContent=request.getParameter("text");
      out.println(textContent);
   %>
<BR><BR>客户端提交的所有参数的名字(request.getParameterNames()):
   <% Enumeration enum1=request.getParameterNames();
      while(enum1.hasMoreElements())
         {String s=(String)enum1.nextElement();
          out.println(s);
         }
   %>
<BR><BR>客户使用的协议是(request.getProtocol()):
   <% String protocol=request.getProtocol();
      out.println(protocol);
   %>
<BR><BR>获取发出请求字符串的客户端地址(request.getRequestURI()):
   <% String url=request.getRequestURI();
      out.println(url);
   %>
<BR><BR>获取客户的IP地址(request.getRemoteAddr()):
   <% String  IP=request.getRemoteAddr();
      out.println(IP);
   %>
<BR><BR>获取客户机的名称(request.getRemoteHost()):
   <% String clientName=request.getRemoteHost();
      out.println(clientName);
   %>
<BR><BR>获取服务器的名称(request.getServerName()):
   <% String serverName=request.getServerName();
      out.println(serverName);
   %>
<BR><BR>获取接受客户提交信息的页面(request.getServletPath()):
   <% String path=request.getServletPath();
      out.println(path);
   %>
<BR><BR>获取服务器的端口号(request.getServerPort()):
   <% int serverPort=request.getServerPort();
      out.println(serverPort);
   %>
</BODY>
</HTML>
```

运行结果如图4-11、图4-12所示。

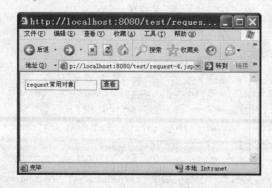

图 4-11　request 对象常用方法的使用

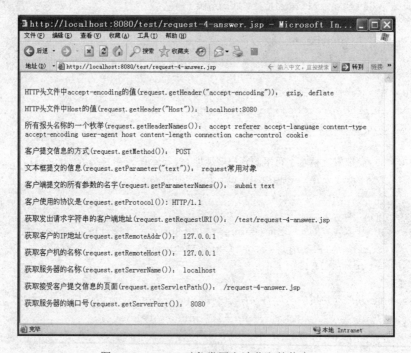

图 4-12　request 对象常用方法获取的信息

除了以上常用的方法，request 对象还提供了其他的方法供用户使用，具体内容读者可参考 J2EE API 的 javax.servlet.http 包中的 HttpServletRequest 类。

## 4.3　response 对象

response 对象对客户的请求做出动态的响应，用来向客户端发送数据，主要是服务器端去设置客户端的某些内容。与 request 相联系的是 HttpServletReponse 类。

### 4.3.1　动态响应 contentType 属性

当一个用户访问一个 JSP 页面时，如果该页面的 contentType 属性是 text/html，那么 JSP 引擎将按照这种属性值作出反应。可以使用 response 对象的 setContentType(String s)方法来改变 contentType 的属性值，参数 s 是内容类型（MIME）的字符串，可以设置为 text/html、application/x-msexcel、application/msword、image/GIF 等。

例 4-8：将当前页面保存成 Word 文档。

```
<%@ page contentType="text/html;charset=GB2312" %>
<HTML>
<BODY>
<P>response 对象的 setContentType 方法
<P>点击 yes 可将当前页面保存为 word 文档
<FORM action="" method="get" name=form>
    <INPUT TYPE="submit" value="yes" name="submit">
</FORM>
<% String str=request.getParameter("submit");
   if(str==null)
      {str="";}
   if(str.equals("yes"))
      {response.setContentType("application/msword;charset=GB2312");}
%>
</BODY>
</HTML>
```

运行结果如图 4-13 所示。

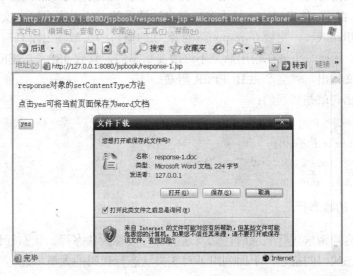

图 4-13　response 对象 setContentType 方法改变 contentType 的属性

### 4.3.2　response 实现网页的自动刷新

response .setheader()方法可以实现网页的自动刷新。

例 4-9：每一秒钟刷新一次页面来显示系统时间。

```
<%@page language="Java"
contentType="text/html;charset=gb2312"
import="java.util.*"
%>
<HTML>
<HEAD>
```

```
<TITLE> response 刷新页面实例</TITLE>
</HEAD>
<BODY>
```

现在的时间是：

```
<%
response.setHeader("refresh","1");
out.println(new Date().toString());
%>
</BODY>
</HTML>
```

### 4.3.3 response 重定向

response 对象可以实现页面的重定向，通过它的 sendRedirect(URL url)方法来实现。该方法通过修改 HTTP 协议的 HEADER 部分,对浏览器下达重定向指令,使浏览器显示重定向网页的内容。

使用 sendRedirect 时，服务器端先响应客户端一个状态码（通常是 302），告诉客户端应该向 location 报头指定的 URL（可以是相对路径）重新发送请求；然后客户端按照指示自动进行第二次请求。由于 sendRedirect 是由客户端自动请求的，所以客户第一次的请求数据就不能得到保存,sendRedirect 一定要在 response 对象,别的语言的头域输出没有输出正文时才能使用，因为按 W3C 标准，头域在输出后是不许改变状态的，只有 JSP 和 SERVLET 可以改变它，而一旦有正文输出，就会产生 BUFFER 阻塞。

例如下面的语句就是错误的：

```
<%
out.println("sss");
response.sendRedirect("url");
%>
```

### 4.3.4 response 的状态行

当服务器对客户请求进行响应时，它发送的首行被称做状态行。状态行包括 3 位数字的状态代码和对状态代码的描述。下面列出了对 5 类状态行代码的大概描述。

- 1yy（1 开头的 3 位数）：主要是实验性质的。
- 2yy：用来表明请求成功。例如，状态代码 204 可以表明已成功取得了请求的页面,但没有新信息。
- 3yy：用来表明在请求满足之前应采取进一步的行动。
- 4yy：当浏览器作出无法满足的请求时，返回该状态行代码。
- 5yy：用来表示服务器出现的问题。例如，500 说明服务器内部发生错误。

例 4-10：response-3.jsp 显示一个带有超级链接的页面，response-3-answer.jsp 显示了设置了状态行的错误页面。

```
response-3.jsp
<%@ page contentType="text/html; charset=gb2312" language="java" import=
            "java.sql.*" errorPage="" %>
```

```
<head>
<meta http-equiv="Content-Type" content="text/html; charset=gb2312" />
<title>无标题文档</title>
</head>
<body>
<a href="response-3-answer.jsp">点击此链接可以进入状态行的测试页面</a>
</body>
</html>
```

response-3-answer.jsp

```
<%@ page contentType="text/html;charset=GB2312" %>
<HTML>
<BODY>
 <%
    response.setStatus(500);
  %>
</BODY>
</HTML>
```

运行结果如图 4-14、图 4-15 所示。

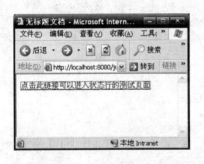

图 4-14 带有超级链接的普通页面

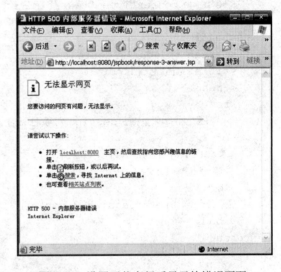

图 4-15 设置了状态行后显示的错误页面

## 4.4 session 对象

session,中文经常翻译为会话,其本来的含义是指有始有终的一系列动作/消息,比如打电话时从拿起电话拨号到挂断电话这中间的一系列过程可以称之为一个 session。

HTTP 协议本身是无状态的,这与 HTTP 协议本来的目的是相符的,客户端只需要简单地向服务器请求下载某些文件,无论是客户端还是服务器都没有必要记录彼此过去的行为,每一次请求之间都是独立的。HTTP 的无状态使得无法知道每次链接是否是同一个客户端的行为。而 session 和 Cookie 就是为了解决 HTTP 协议无状态的缺陷所作出的努力。

当一个客户端访问服务器时,可能有时会反复刷新一个页面,或者是在几个页面之间来回跳转等，这时服务器可以通过 session 对象存放在客户端的 Cookie 里的内容来知道这是同一个客户端发送的请求。

### 4.4.1 session 对象的 ID

当客户端访问服务器上的 JSP 页面时，服务器首先检查这个客户端的请求里是否已包含了一个 session 标识-称为 session ID，如果已包含一个 session ID 则说明以前已经为此客户端创建过 session，服务器就按照 session ID 把这个 session 检索出来使用（如果检索不到，可能会新建一个）；如果客户端请求不包含 session ID，则为此客户端创建一个 session 并且生成一个与此 session 相关联的 session ID，这个 session ID 将同时被发送给客户端，存放在客户端的 Cookie 中。这样，session 对象和客户之间建立了一个一一对应的关系，每个不同的客户对应着不同的 session 对象。对于某个客户来说，只要不关闭浏览器，就一直是使用的初始的 session 对象，也就是说一直具有同一个 session 的 ID，直到关闭浏览器取消该 session 对象。

**例 4-11**：session-1.jsp 和 session-2.jsp 可以显示 session 对象 ID，在不关闭浏览器的情况下，该 ID 的值是一样的。

```
session-1.jsp
<%@ page contentType="text/html; charset=gb2312" language="java" import=
                 "java.sql.*" errorPage="" %>
<head>
<meta http-equiv="Content-Type" content="text/html; charset=gb2312" />
<title>无标题文档</title>
</head>
<body>
<h3>这是第一个页面，session 对象对应的 ID 是：
<br />
  <%String id=session.getId();%>
  <%=id%></h3>
 <br />
  <h3><a href="session-2.jsp">链接到第二个页面</a>  </h3>
</body>
</html>

session-2.jsp
<%@ page contentType="text/html; charset=gb2312" language="java" import=
                 "java.sql.*" errorPage="" %>
<head>
<meta http-equiv="Content-Type" content="text/html; charset=gb2312" />
<title>无标题文档</title>
</head>
<body>
<h3>这是第二个页面，session 对象对应的 ID 是：
<br />
  <%String id=session.getId();%>
<%=id%></h3>
  <h3><a href="session-1.jsp">链接到第一个页面</a>  </h3>
</body>
</html>
```

运行结果如图 4-16、图 4-17 所示。

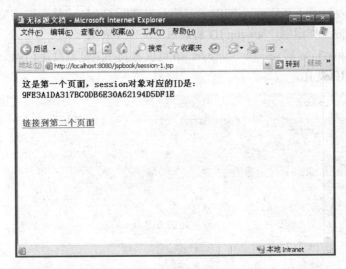

图 4-16　显示第一个页面时使用的 ID

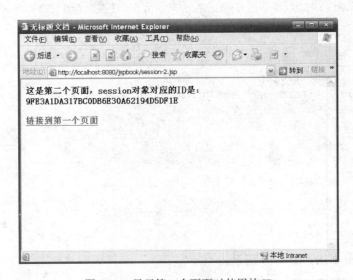

图 4-17　显示第二个页面时使用的 ID

### 4.4.2　session 对象与 URL 重写

因为 session ID 在创建的同时被发送给客户端，存放在客户端的 Cookie 中，所以，只有在客户的浏览器支持 Cookie 的情况下，session 对象才能和客户建立起一一对应的联系。例如，在上述例子中，如果禁用了浏览器中的 Cookie，那么每个页面使用的 session 的 ID 就各不相同，也就是说每个客户对应了多个 session 对象。

解决这个问题的办法就是通过 URL 重写。URL 重写功能其实就是对 URL 进行一个编码加密，客户端看不到真实的 URL。但是它还有一个重要的功能就是每次进行完 URL 重写之后都会将 session ID 放置到 URL 中，这样每次服务器在获取相应的 session 时需要的 session ID 可以从 URL 中找到。而且 URL 还进行了加密，增强了安全性。

把 session ID 加到一个连接可以使用一对方法来做：response.encodeURL() 使 URL 包含

session ID。在需要使用重定向时，可以使用 response.encodeRedirectURL()来对 URL 进行编码。encodeURL()及 encodeRedirectedURL()方法首先判断 Cookies 是否被浏览器支持，如果支持，则参数 URL 被原样返回，session ID 将通过 Cookies 来维持。

**例 4-12**：session-3.jsp 和 session-3-answer.jsp 可以看到在浏览器设置 Cookie 和不设置 Cookie 时 session 对象的 ID 变化。

```jsp
session-3.jsp
<%@ page contentType="text/html;charset=GB2312" %>
<HTML>
<BODY>
<P> 本页面的session 对象的 ID 是：
<br>
  <% String s=session.getId();
     String str=response.encodeURL("session-3-answer.jsp");
  %>
 <%=s%>
 <BR> <BR> <BR>
    <a href=session-3-answer.jsp>转向另外一个页面查看它的 session ID</a>
</BODY>
</HTML>


session-3-answer.jsp
<%@ page contentType="text/html;charset=GB2312" %>
<HTML>
<BODY>
    <% String s=session.getId();
    %>
<P> answer 页面的session 对象的 ID 是：
<br>    <%=s%>
</BODY>
</HTML>
```

运行结果如图 4-18、图 4-19 所示。

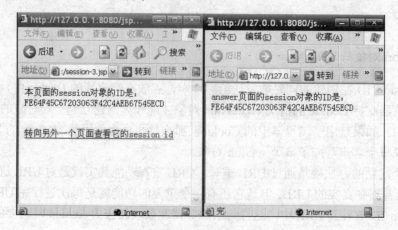

图 4-18 设置 Cookie 时的 session ID 显示

图 4-19 不使用 Cookie 时的 session ID 显示

### 4.4.3 session 对象的常用方法

session 对象指的是客户端与服务器的一次会话,从客户连到服务器的一个 WebApplication 开始,直到客户端与服务器断开连接为止,它是 HttpSession 类的实例。

- session.getId:获取 session 的 ID。
- public boolean isNew():判断当前的 session 是否是新建立的 session。
- public void setAttribute(String name,Object obj):设置属性,该方法可以将 obj 添加到 session 对象中,并为其指定一个名字。
- public Object getAttribute(String name):获得属性,即返回 session 对象中指定名字绑定的对象。
- public void removeAttribute(String name):删除属性,删除指定名字的绑定。
- public long getCreationTime():得到 session 的创建时间,此方法返回 long 类型,通过 Date 类可以取得一个完整时间。
- public long getLastAccessedTime():获取当前 session 对象最后一次被操作的时间,单位是毫秒。
- public void invalidate():使当前 session 失效。

例 4-13:通过 setAttribute 方法和 getAttribute 方法来判断用户是否进行了正确的登录,用户名和密码都是 aaa,输入正确时才能进入页面。

```
session-4.jsp
<%@ page contentType="text/html; charset=gb2312" language="java" import=
                "java.sql.*" errorPage="" %>
<html xmlns="http://www.w3.org/1999/xhtml">
<head>
<meta http-equiv="Content-Type" content="text/html; charset=gb2312" />
<title>无标题文档</title>
</head>
<body>
    <form action="session-4-answer1.jsp" method="post">
        用户名:<Input type="text" name="name" /><br><br>
        密码:<Input type="password" name="password" />
```

```
        <br><br>
        <label>
            <input type="submit" name="Submit" value="提交">
        </label>
    </form>
</body>
</html>
```

session-4-answer1.jsp
```
<%@ page contentType="text/html; charset=gb2312" language="java" import=
                "java.sql.*" errorPage="" %>
<html xmlns="http://www.w3.org/1999/xhtml">
<head>
<meta http-equiv="Content-Type" content="text/html; charset=gb2312" />
<title>无标题文档</title>
</head>
<body>
<%
        if((!(request.getParameter("name")).equals(""))&&(!(request.
            getParameter("password")).equals(""))){
    String name= request.getParameter("name");
    String password= request.getParameter("password");
    if(name.equals("aaa")&&password.equals("aaa")){//表示用户存在且密码正确
        session.setAttribute("flag",name);
        response.sendRedirect("session-4-answer2.jsp");
    }
    else{out.println("用户名或密码错误,登录失败");
    }
    }
else{   session.setAttribute("flag",null);
    response.sendRedirect("session-4-answer2.jsp");
    }
    %>
</body>
</html>
```

session-4-answer2.jsp
```
<%@ page contentType="text/html; charset=gb2312" language="java" import=
                "java.sql.*" errorPage="" %>
<html xmlns="http://www.w3.org/1999/xhtml">
<head>
<meta http-equiv="Content-Type" content="text/html; charset=gb2312" />
<title>无标题文档</title>
</head>
<body>
 <!--用户必须登录后才能访问此页面-->
  <% if((session.getAttribute("flag")) != null){//session被设置,正常登录过
     out.println("欢迎进入本页面!"); }
    else{//对于没有登录的用户,输出错误信息
```

```
            out.println("对不起,您还没有输入用户名和密码进行登录"); }
        %>
</body>
</html>
```

运行结果如图 4-20～图 4-22 所示。

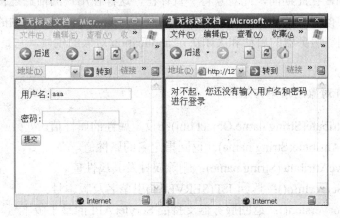

图 4-20 当用户名或密码没输入时

图 4-21 当用户名或密码输入错误时

图 4-22 当用户名和密码输入正确时

## 4.5 application 对象

application 对象实现了用户间数据的共享，可存放全局变量。它开始于服务器的启动，直到服务器的关闭，在此期间，此对象将一直存在，这样在用户的前后连接或不同用户之间的连接中，可以对此对象的同一属性进行操作，在任何地方对此对象属性的操作都将影响其他用户对此的访问。服务器的启动和关闭决定 application 对象的生命。它是 ServletContext 类的实例。

### 4.5.1 application 对象的常用方法

① void setAttribute(String name,Object obj)：设定属性的属性值。
② Object getAttribute(String name)：返回指定名的属性值。
③ void removeAttribute(String name)：删除属性及其属性值。
④ String getServerInfo()：返回 JSP(SERVLET)引擎名及版本号。
⑤ int getMinorVersion()：返回服务器支持的 Servlet API 的最小版本号。
⑥ int getMajorVersion()：返回服务器支持的 Servlet API 的最大版本号。
⑦ String getRealPath(String path)：返回一虚拟路径的真实路径。
⑧ URL getResource(String path)：返回指定资源（文件及目录）的 URL 路径。

例 4-14：application 对象的常用方法的使用。

```jsp
<%@ page contentType="text/html; charset=gb2312" language="java" import=
                "java.sql.*" errorPage="" %>
<html xmlns="http://www.w3.org/1999/xhtml">
<head>
<meta http-equiv="Content-Type" content="text/html; charset=gb2312" />
<title>无标题文档</title>
</head>
<body>
JSP(SERVLET)引擎名及版本号:<%=application.getServerInfo()%>  <br>
返回/application1.jsp 虚拟路径的真实路径:<%=application.getRealPath
    ("/application-1.jsp")%>  <br>
服务器支持的 Servlet API 的大版本号:<%=application.getMajorVersion()%>  <br>
服务器支持的 Servlet API 的小版本号:<%=application.getMinorVersion()%>  <br>
指定资源(文件及目录)的 URL 路径:<%=application.getResource("/application1.
    jsp")%>  <br>
<% application.setAttribute("name","application 对象的 setAttribute 方法的使用");
out.println(application.getAttribute("name"));
application.removeAttribute("name");
out.println(application.getAttribute("name"));
%>
<br>
<%
Object count=application.getAttribute("count");
int i=0;
if(count==null) {application.setAttribute("count","1"); }
```

```
else{
    application.setAttribute("count",i);
    i=Integer.parseInt(count.toString());
    application.setAttribute("count",i+1);}
%>
你是第<%=application.getAttribute("count")%>位访问者
</body>
</html>
```

运行结果如图 4-23 所示。

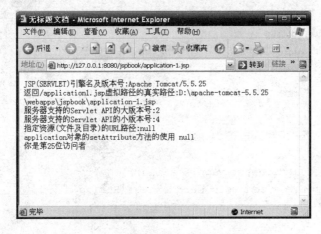

图 4-23　application 对象常用方法的使用

### 4.5.2　用 application 对象制作留言板

**例 4-15**：制作简易的留言版页面，用户可以在上面添加留言信息，然后进行提交，也可以查看其他的留言信息。文件 application-2.jsp 通过表单让用户输入信息。文件 application-3.jsp 接收用户的输入信息，并调用自定义函数 sendMessage()将信息存储到向量 v 中。由于可能有些服务器不直接支持使用 application 对象，所以例子 application-3.jsp 中先使用 ServletContext 类声明这个对象，然后使用 getServletContext()方法对 application 对象进行初始化。在文件 application-4.jsp 中，使用 StringTokenizer 类对向量 v 中存储的内容通过循环进行输出。

```
application-2.jsp
<%@ page contentType="text/html;charset=GB2312" %>
<HTML>
<BODY>
 <FORM action="application-3.jsp" method="post" name="form">
    <P>请输入您的姓名：
    <INPUT type="text" name="name">
    <P>请输入您的电子信箱：
      <input type="text" name="email">
    <P>请输入您要留言的标题：
      <INPUT type="text" name="title">
    <P>请输入您的留言：
    <textarea name="message" rows="8" cols=30></textarea>
```

```
    <P><BR>
      <INPUT type="submit" value="提交" name="submit">
    <P>
      <input type="reset" name="reset" value="重填">
    </FORM>
 <FORM action="application-4.jsp" method="post" name="form1">
    <INPUT type="submit" value="查看留言板" name="look">
 </FORM>
</BODY>
</HTML>

application-3.jsp
<%@ page contentType="text/html;charset=GB2312" %>
<%@ page import="java.util.*" %>
<HTML>
<BODY>
    <p>
      <%! Vector v=new Vector();
         int i=0;
         ServletContext application;
         synchronized void sendMessage(String s)
          { application=getServletContext();;
            i++;
            v.add("No."+i+":"+"#"+s);
            application.setAttribute("Message",v);
          }
      %>
       <% request.setCharacterEncoding("GBK");
         String name=request.getParameter("name");
         String email=request.getParameter("email");
         String title=request.getParameter("title");
         String message=request.getParameter("message");
           if(name==null)
             {name="guest";
             }
           if(email==null)
             {email="没有输入电子信箱";
             }
           if(title==null)
             {title="没有输入标题";
             }
           if(message==null)
             {message="没有留言信息";
             }
         String s="NAME:"+name+"#"+"E-MAIL:"+email+"#"+"TITLE:"+title
              +"#"+"MESSAGE:"+message;
         sendMessage(s);
         out.print("您的信息已经提交! ");
```

```
      %>
    </p>
    <p><A HREF="application-2.jsp" >返回</A></p>
</BODY>
</HTML>


application-4.jsp
<%@ page contentType="text/html;charset=GB2312" %>
<%@ page import="java.util.*" %>
<HTML>
<BODY>
    <%Vector v=(Vector)application.getAttribute("Message");
      for(int i=0;i<v.size();i++)
        { String message=(String)v.elementAt(i);
          StringTokenizer liuyan=new StringTokenizer(message,"#");
          while(liuyan.hasMoreTokens())
            { String str=liuyan.nextToken();
              out.print("<BR>"+str);
            }
        }
    %>
</BODY>
</HTML>
```

运行结果如图 4-24～图 4-26 所示。

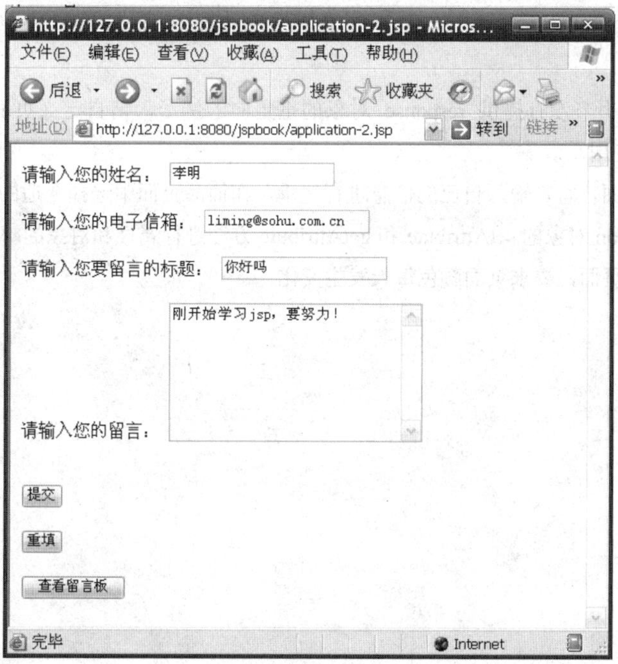

图 4-24　application-2.jsp 运行结果页面

图 4-25  application-3.jsp 运行结果页面

图 4-26  application-4.jsp 查看留言板页面

## 4.6 小  结

本章主要讲解了 JSP 内置的 5 个对象及其常用的使用方法，JSP 除了提供本章讲到的 5 个内置对象外，还有 pageContext(javax.servlet.jsp.PageContext)、config(javax.servlet.ServletConfig)、exception（java.lang.Throwable）、page（javax.servlet.jsp.HttpJspPage）4 个内置对象。pageContext 对象存储本 JSP 页面相关信息，如属性、内建对象等；config 对象用来存放 Servlet 初始的数据结构；exception 错误对象，只有在 JSP 页面的 page 指令中指定 isErrorPage="true"后，才可以在本页面使用 exception 对象；page 对象代表 JSP 对象本身，或者说代表编译后的 servlet 对象。读者可以参考相应的 api 文档来研究其他几个内置对象。

## 4.7 习  题

1．本章中 4.3.3 小节没有提供 response 对象的重定向的例子，请自定义页面内容，对 response 的 sendRedirect 方法进行使用。

2．建立购物车页面，客户输入自己的信息进行登录，在商品页面中选择合适的商品选择购买，然后到结账处结账。使用 session 对象的 setAttribute 和 getAttribute 方法进行属性和对象的绑定。

3．设计一个 JSP 页面，要求页面颜色每天发生变化。

# 第 5 章 Servlet

了解 Servlet 的原理，掌握 Servlet 的编译、安装与运行的步骤，掌握 Servlet 的生命周期以及其基本方法，掌握 Servlet 与 JSP 的数据交互方法，掌握 Servlet 的高级用法。

## 5.1 Servlet 概述

Servlet 是使用 Java Servlet 应用程序设计接口（API）及相关类和方法的 Java 程序，是一个专门用于编写网络服务器应用程序的 Java 组件。除了 Java Servlet API，Servlet 还可以使用用以扩展和添加到 API 的 Java 类软件包。Servlet 在启用 Java 的 Web 服务器上或应用服务器上运行并扩展了该服务器的能力。Java Servlet 对于 Web 服务器就好像 Java applet 对于 Web 浏览器。Servlet 装入 Web 服务器并在 Web 服务器内执行，而 applet 装入 Web 浏览器并在 Web 浏览器内执行。

### 5.1.1 Servlet 简介和优点

自 1997 年 3 月 Sun Microsystems 公司所组成的 JavaSoft 部门将 Servlet API 定案以来，推出了 Servlet API 1.0，就当时功能来说，Servlet 所提供的功能包含了当时的 CGI 与 Netscape Server API（NSAPI）之类产品的功能。发展至今，它依旧是一个具有跨平台特性、100% Pure Java 的 Server-Side 程序，Servlet 不只限定于 HTTP 协议，开发人员可以利用 Servlet 自定义或延伸任何支持 Java 的 Server，包括 Web Server、Mail Server、Ftp Server、Application Server 或任何自定义的 Server。

Server 有以下优点：

• **可移植性**，Servlet 是利用 Java 语言来开发的，因此，延续了 Java 在跨平台上的表现，不论 Server 的操作系统是什么，如 Windows、Linux、Solaris、HP-UX 等，都能够将我们写好的 Servlet 程序放在这些操作系统上执行，借助 Servlet 的优势，就可以真正达到 Write Once，Serve Anywhere 的境界。Servlet 是在 Server 端执行的，所以，程序员只要专心开发，能在实际应用的平台环境下测试无误即可。除非从事做 Servlet Container 的公司，否则不须担心写出来的 Servlet 是否能在所有的 Java Server 平台上执行。

• **强大的功能**，Servlet 能够完全发挥 Java API 的威力，包括网络和 URL 存取、多线程、影像处理、RMI（Remote Method Invocation）、分布式服务器组件、对象序列化等。若想写个网络目录查询程序，则可以利用 JNDI API，相连接数据库可以用 JDBC，有这些强大功能的 API 做后盾，Servlet 更能发挥其优势。

• **高性能**，Servlet 在加载执行后，其对象实体通常会一直停留在 Server 的内存中，若有请求发生时，服务器再调用 Servlet 来服务，假若收到相同服务的请求时，Servlet 会利用不同的线程来处理，不像 CGI 程序必须产生许多进程来处理数据。在性能表现上，它大大超过 CGI 程序。Servlet 在执行时，不是一直停留在内存中，服务器会自动将停留时间过长一直没有执行的 Servlet 从内存中移除，不过有时候也可以自行写程序来控制，至于停留时间长短通常和选用的服务器有关。

- **安全性**，Servlet 也有类型检查的特性，并且利用 Java 的垃圾回收与没有指针的设计，使得 Servlet 避免内存管理的问题。在 Java 的异常处理机制下，Servlet 能够安全地处理各种错误，不会因为发生程序上逻辑错误而导致整体服务器系统的崩溃。例如，在某个 Servlet 发生除以零或其他不合法的运算时，会抛出一个异常让服务器处理，如记录在 Log 日志中。

### 5.1.2 Servlet 与 JSP 的关系

Servlet 是用 Java 编写的 Server 端程序，它与协议和平台无关。Servlet 运行于 Java-enabled Web Server 中。Java Servlet 可以动态地扩展 Server 的能力，并采用请求-响应模式提供 Web 服务。JSP 是 Java Server Page 的缩写，是 Sun 公司出品的 Web 开发语言，是一种动态网页技术标准。它在 HTML 代码中插入 JSP 标记（tag）及 Java 程序片段（Scriptlet），构成 JSP 页面。当客户端请求 JSP 文件时，Web 服务器执行该 JSP 文件，然后以 HTML 的格式返回给客户。同时由于它的跨平台性，愈来愈受到广泛的应用。Servlet 与 JSP 之间的交互为开发 Web 服务提供了优秀的解决方案。

Servlet 与 JSP 之间有着紧密的联系。JSP 只是构建在 Servlet 之上的高层次的动态网页标准，因此，从概念上讲，相对 Servlet 而言，JSP 并没有什么新的东西，可以说在概念上 JSP 跟 Servlet 是完全一样的，只不过在实现方法上稍有不同。在 JSP 执行时，JSP 引擎会按照 JSP 的语法，将 JSP 文件转换成 Servlet 代码源文件，接着 Servlet 会被编译成 Java 可执行字节码（bytecode），并以一般的 Servlet 方式载入执行。

JSP 与 Servlet 之间的主要差异在于，JSP 提供了一套简单的标签，和 HTML 融合得比较好，使不了解 Servlet 的人也可以做出动态网页来。对于 Java 语言不熟悉的人会觉得 JSP 开发比较方便。JSP 修改后可以立即看到结果，不需要手工编译，JSP 引擎会来做这些工作；而 Servlet 却需要编译，重新启动 Servlet 引擎等一系列动作。

JSP 语法简单，可以很方便地嵌入 HTML 之中，很容易加入动态的部分，很方便地输出 HTML。而在 Servlet 中输出 HTML 却需要调用特定的方法，对于引号之类的字符也要做特殊的处理，加在复杂的 HTML 页面中作为动态部分，比起 JSP 来说是比较困难的。但是在 JSP 中，HTML 与程序代码混杂在一起，而 Servlet 的逻辑型性要更强一些。

从表面上看，JSP 页面已经不再需要 Java 类，似乎完全脱离了 Java 面向对象的特征。事实上，JSP 是 Servlet 的一种特殊形式，每个 JSP 页面就是一个 Servlet 实例——JSP 页面由系统编译成 Servlet，Servlet 再负责响应用户请求。JSP 其实也是 Servlet 的一种简化，使用 JSP 时，其实还是使用 Servlet，因为 Web 应用中的每个 JSP 页面都会由 Servlet 容器生成对应的 Servlet。对于 Tomcat 而言，JSP 页面生成的 Servlet 放在 work 路径对应的 Web 应用下。

### 5.1.3 JSP 文件编译过程

Web 容器（如 Tomcat）处理 JSP 文件请求的执行过程主要包括以下 4 个部分：
- 客户端发出 Request 请求。
- JSP 容器将 JSP 转译成 Servlet 的源代码。
- 将产生的 Servlet 源代码经过编译后，并加载到内存执行。
- 把结果 Response（响应）至客户端。

很多人都会认为 JSP 的执行性能会和 Servlet 相差很多，其实执行性能上的差别只在第一次的执行。因为 JSP 在执行第一次后，会被编译成 Servlet 的类文件，即.class，当再重复调用执行时，就直接执行第一次所产生的 Servlet，而不再重新把 JSP 编译成 Servlet。因此，除了第一次的编译会花较久的时间之外，之后 JSP 和 Servlet 的执行速度就几乎相同了。

在执行 JSP 网页时，通常可以分为两个时期：转译时期（Translation Time）和请求时期（Request Time）。

转译时期：JSP 网页转移成 Servlet 类。

请求时期：Servlet 类执行后，响应结果至客户端。

转译期间做了以下两件事情。

转译时期：将 JSP 网页转移为 Servlet 源代码 .java。

编译时期：将 Servlet 源代码 .java 编译成 Servlet 类 .class。

当 JSP 网页在执行时，JSP 容器会做检查工作，如果发现 JSP 网页有更新修改时，JSP 容器才会再次编译 JSP 成 Servlet；如果 JSP 没有更新时，就直接执行前面所产生的 Servlet。

例 5-1：以清单 5-1 所示 showdate.jsp 程序为例说明整个过程。

清单 5-1　showdate.jsp 程序

```
<%@ page language="java" contentType="text/html;charset=gb2312" import=
          "java.text.*,java.util.*;"%>
<html>
<head>
<title>Show time</title>
</head>
<body>
    Hello :
    <%
        SimpleDateFormat format = new SimpleDateFormat("yyyy/MM/dd");
        String str = format.format(new Date());
    %>
    <%=str %>
</body>
</html>
```

当部署好 showdate.jsp 之后，启动 Tomcat 服务器。

（1）在浏览器中输入配置好的路径请求，showdate.jsp 页面。

（2）Tomcat 服务器将 showdate.jsp 转译成 showdate_jsp.java 源文件。

（3）同时将 showdate_jsp.java 源文件编译成 showdate_jsp.class。

（4）编译执行 showdate_jsp.class 类，处理请求，返回响应，容器将生成的页面返回给客户端显示。

转移成的 Java 源文件。showdate_jsp.java 如清单 5-2 所示。

清单 5-2　showdate.jsp 程序

```
package org.apache.jsp;

import javax.servlet.*;
import javax.servlet.http.*;
import javax.servlet.jsp.*;
```

```java
      import java.text.*;
      import java.util.*;

public final class showdate_jsp extends org.apache.jasper.runtime.HttpJspBase
    implements org.apache.jasper.runtime.JspSourceDependent {

  private static java.util.List _jspx_dependants;

  public Object getDependants() {
    return _jspx_dependants;
  }

  public void _jspService(HttpServletRequest request, HttpServletResponse response)
        throws java.io.IOException, ServletException {
    JspFactory _jspxFactory = null;
    PageContext pageContext = null;
    HttpSession session = null;
    ServletContext application = null;
    ServletConfig config = null;
    JspWriter out = null;
    Object page = this;
    JspWriter _jspx_out = null;
    PageContext _jspx_page_context = null;

    try {
      _jspxFactory = JspFactory.getDefaultFactory();
      response.setContentType("text/html;charset=gb2312");
      pageContext = _jspxFactory.getPageContext(this, request, response,
            null, true, 8192, true);
      _jspx_page_context = pageContext;
      application = pageContext.getServletContext();
      config = pageContext.getServletConfig();
      session = pageContext.getSession();
      out = pageContext.getOut();
      _jspx_out = out;

      out.write("\r\n");
      out.write("<html>\r\n");
      out.write("<head>\r\n");
      out.write("<title>Show time</title>\r\n");
      out.write("</head>\r\n");
      out.write("<body> \r\n");
      out.write("\tHello : \r\n");
      out.write("\t");

          SimpleDateFormat format = new SimpleDateFormat("yyyy/MM/dd");
          String str = format.format(new Date());
```

```
        out.write("\r\n");
        out.write("\t ");
        out.print(str );
        out.write("\r\n");
        out.write("</body>\r\n");
        out.write("</html>");
      } catch (Throwable t) {

        if (!(t instanceof SkipPageException)){
          out = _jspx_out;
          if (out != null && out.getBufferSize() != 0)
            out.clearBuffer();
          if (_jspx_page_context != null) _jspx_page_context.
             handlePageException(t);
        }

      } finally {

        if (_jspxFactory != null) _jspxFactory.releasePageContext
           (_jspx_page_context);
      }
    }
  }
```

当 JSP 页面被转译成 Servlet 时，内容主要包含三个部分：

```
public void _jspInit(){...} //当JSP网页一开始执行时,最先执行此方法,执行初始化工作
public void _jspDestory(){...}  //JSP网页最后执行的方法
public void _jspService(HttpServletRequest request, HttpServletResponse
                        response)
    throws java.io.IOException, ServletException {
                            //JSP网页中最主要的程序都是在此执行
```

将 showdate.jsp 和 showdate_jsp.java 做一个简单对比。

第一部分，页面属性的对比。

showdate.jsp 页面中代码：

```
<%@ page language="java" contentType="text/html;charset=gb2312" %>
```

showdate_jsp.java 中等价代码：

```
response.setContentType("text/html;charset=gb2312");
//通过 response 响应设置返回客户端的页面属性
```

第二部分，HTML 标签：

showdate.jsp 页面中代码：

```
<html>
<head>
<title>Show time</title>
</head>
```

```
        ...
        </html>
        showdate_jsp.java 中等价代码：
        out.write("\r\n");
        out.write("<html>\r\n");
        out.write("<head>\r\n");
        out.write("<title>Show time</title>\r\n");
        out.write("</head>\r\n");
        out.write("<body> \r\n");
        out.write("\tHello : \r\n");
        out.write("\t");
        //通过 out 对象 向客户端写 HTML 标签
```

第三部分，声明的对象。
showdate.jsp 页面中代码：

```
<%
        SimpleDateFormat format = new SimpleDateFormat("yyyy/MM/dd");
        String str = format.format(new Date());
%>
```

showdate_jsp.java 中等价代码（在 _jspService 方法中声明的局部变量）：

```
SimpleDateFormat format = new SimpleDateFormat("yyyy/MM/dd");
String str = format.format(new Date());
```

第四部分：表达式。
showdate.jsp 页面中代码：

```
<%=str %>
```

showdate_jsp.java 中等价代码：

```
out.print(str );    //写 即打印 str 变量的值
```

### 5.1.4 HTTP 基础知识

用户的请求和 Web 应用程序的相应需要通过 Internet 从一台计算机发送到另一台计算机或服务器，使用超文本传输协议 HTTP。HTTP 是互联网上应用最为广泛的一种网络协议，是一个客户端和服务器端请求和应答的标准。客户端是终端用户，服务器端是网站。通过使用 Web 浏览器等工具，客户端发起一个到服务器上指定端口的 HTTP 请求。应答的服务器上存储着一些资源，比如 HTML 文件和图像。HTTP 协议并没有规定必须使用它和基于它支持的层。事实上，HTTP 可以在任何其他互联网协议上或者其他网络上实现。HTTP 只假定其下层协议提供可靠的传输，任何能够提供这种保证的协议都可以被其使用。

1. HTTP 请求、响应和头信息

客户端发送的请求消息为字符流，由请求行（包括方法、统一资源标识符 URI 和 HTTP 协议版本）和头信息组成。

清单 5-3 是一个合法的 HTTP 请求消息的例子。通过这个例子可大概了解 HTTP 请求消息结构。

清单 5-3　一个典型的 HTTP 请求消息

```
GET /showdate.jsp HTTP/1.1
Host: localhost:8080
User-Agent: Mozilla/5.0 (Windows; U; Windows NT 6.0; zh-CN; rv:1.9.0.11)
Accept: text/html,application/xhtml+xml,application/xml,*/*
Accept-Language: zh-cn
Accept-Charset: gb2312,utf-8
```

这里，请求行指定了用 GET 方法，访问一个名为/showdate.jsp 的资源，并指定请求使用 HTTP/1.1 协议版本。

请求的方法不仅只有 GET，还有 POST、HEAD、OPTIONS、DELETE、PUT、TRACE。其中，常用的是 GET 和 POST 方法，后续内容会详细讲解，其余的各个方法简单了解即可。

Host 首部可以像服务器通知 URL 中所用的主机名。

User-Agent 首部包含了发出请求的浏览器的类型相关信息。服务器可以使用此信息向不同类型的浏览器发送不同类型的响应。

Accept 首部提供了浏览器所接受语言和文件格式的有关信息。这些首部可以用于针对浏览器的功能和用户的首选项（如使用某种所支持的图像格式和首选语言）对响应加以调整。

请求信息由服务器进一步处理，并生成相应的响应，响应消息由状态行和头信息组成，如代码清单 5-4 所示。

清单 5-4　响应消息

```
HTTP/1.x 200 OK
Server: Apache-Coyote/1.1
Content-Type: text/html
Content-Length: 186
Date: Wed, 17 Jun 2009 00:57:35 GMT
```

第一行为状态行，状态行中的状态码 200 只是已经成功处理请求，因此描述为 OK。

常见的 3 种状态码：404 表示找不到被请求的网页时常遇到的状态码；500 表示服务器内部错误；503 表示服务器超时等。如果需要具体了解详情，请访问http://www.w3c.org网站相关内容。

Server 头部指定服务器软件。

Content-Type 头部指定了文档的 MIME 类型，如果是 text/html 指示格式化的 Web 文档，如果是 text/plain 指示未格式化的文本文档。

Content-length 头部指定文件的大小，以字节来表示。

以上是请求和响应消息中可能包括的少数首部。更加详细的介绍可以查看http://www.w3c.org网页。

2. GET 和 POST 方法的区别

HTTP 请求消息使用 GET 或 POST 方法，以便在 Web 上传输请求。

检索信息时一般用 GET 方法，如检索文档、图表或数据库查询结果。要检索的信息作为字符序列传递，称为查询字符串。因此，传递的数据对客户端是可见的，即将查询字符串附加到 URL 中，但是，对查询字符串的长度有限制，最多 1024 字节。GET 方法是表单默认的方法。

我们用 Google 检索 servlet，可以知道 Google 使用了 GET 方法对用户输入的搜索字符串检索搜索结果，如图 5-1 所示。

图 5-1  GET 方法检索页面

HTTP 定义的另一种请求方法是 POST 方法。使用 POST 发送的数据对客户端是不可见的，且对发送的数据的量没有限制。

下面我们来对比一下 GET 和 POST 方法。
- GET 是从服务器上获取数据；POST 是向服务器传送数据。
- 在客户端，GET 通过 URL 提交数据，数据在 URL 中可见；POST 把数据放在 form 的数据体内提交。
- GET 提交的数据最多只有 1024 字节；POST 提交的数据量无限制。
- 由于使用 GET 时，参数会显示在地址栏上，而 POST 不会，所以，如果这些数据是非敏感数据，那么使用 GET；如果包含敏感数据，为了安全，使用 POST。

## 5.2  Servlet 的编译和运行

Servlet 类的编译需要 Servlet-api.jar 包的支持，如果你希望像编译其他 Java 源文件一样用 javac 命令进行编译的话，只需要将 Servlet-api.jar 配置到系统的环境变量 classpath 之下就可以了；如果使用开发工具编译 Servlet，只需要将 Servlet-api.jar 包加入的开发工具的编译环境中，具体配置方法请参看其相关文档。

### 5.2.1  一个简单的 Servlet 例子

**例 5-2**：通过简单的 HelloWorld Servlet 例子来说明 Servlet 程序的建立，程序代码的功能和作用将在后续内容中讲解。清单 5-5 所示的代码为在浏览器上显示 hello world 的 Servlet 程序 HelloWorld.java。

清单 5-5  显示 Hello World 的 HelloWorld.java 程序

```java
import javax.servlet.*;
import javax.servlet.http.*;


public class HelloWorld extends HttpServlet {

public void init(ServletConfig config) throws ServletException {
```

```java
    super.init(config);
}

public void destroy() {
}

protected void processRequest(HttpServletRequest request,
    HttpServletResponse response)
throws ServletException, java.io.IOException {
try{
response.setContentType("text/html");
java.io.PrintWriter out = response.getWriter();
out.write("<html>\n");
out.write("<head>\n");
out.write("<title>Basic Servlet</title>\n");
out.write("</head>\n");
out.write("<body>\n");
out.write("<h1>Hello world!</h1>\n");
out.write("</body>\n");
out.write("</html>");
out.close();
}catch(Exception e){
throw new ServletException(e);
}
}
protected void doGet(HttpServletRequest request,
HttpServletResponse response)
throws ServletException, java.io.IOException {
    processRequest(request, response);
}
protected void doPost(HttpServletRequest request,
HttpServletResponse response)
throws ServletException, java.io.IOException {
    processRequest(request, response);
}

}
```

### 5.2.2 存放 Servlet 的目录

一个 Java Web 应用由 Web 组件（Servlet、JSP 等）和静态资源（HTML、图片等）构成，其目录结构如图 5-2 所示。

- JSP、静态资源（HTML、图片）位于应用程序目录或其子目录下。
- WEB-INF/classes 目录放置 Servlet 类。
- WEB-INF/lib 目录放置 Java Archive 文件。
- WEB-INF 目录下的 web.xml 文件是部署描述符。

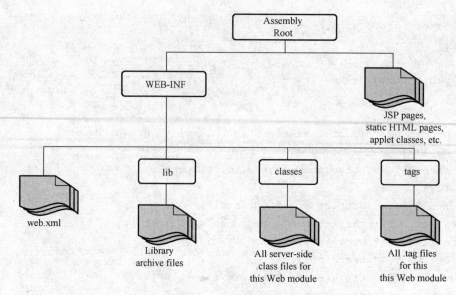

图 5-2　Web 应用的目录结构

### 5.2.3　运行 Servlet

运行清单 5-5 所示的 Servlet 程序需要如下几步操作：

（1）创建 Web 开发目录，例如 HelloWorld，放在 Tomcat 的 webapps 目录下。

（2）编译 Servlet 类 HelloWorld.java（用 javac 编译时需要将 Servlet-api.jar 包加入系统的环境变量中），把生成的 class 文件 HelloWorld.class 放置于 Web 应用/HelloWorld/WEB-INF/classes 目录下。

（3）部署 Web.xml。修改/HelloWorld/WEB-INF 目录下的部署描述符 Web.xml，添加 Servlet 映射，如清单 5-6 所示，黑体部分为注册添加的代码。

由于在运行 Servlet 时，都必须要在 Web.xml 中添加相应的 Servlet 元素，所以简单地把 <servlet>元素的具体属性介绍一下。

- <servlet-name>定义 Servlet 的名字，也就是给 Servlet 类起别名。
- <servlet-class>指定实现这个 Servlet 的类。
- 在<servlet-mapping>元素中，<servlet-name>指定在<servlet>元素中相对应的 Servlet 的名字，<url-pattern>指定访问 Servlet 的相对 URL 路径。

清单 5-6　web.xml

```
<?xml version="1.0" encoding="ISO-8859-1"?>

<web-app xmlns="http://java.sun.com/xml/ns/javaee"
  xmlns:xsi="http://www.w3.org/2001/XMLSchema-instance"
  xsi:schemaLocation="http://java.sun.com/xml/ns/javaee
            http://java.sun.com/xml/ns/javaee/web-app_2_5.xsd"
  version="2.5">
  <display-name>Welcome to Tomcat</display-name>
  <description>
    Welcome to Tomcat
```

```xml
        </description>

    <servlet>
        <servlet-name>hello</servlet-name>
        <servlet-class>HelloWorld</servlet-class>
    </servlet>
    <servlet-mapping>
        <servlet-name>hello</servlet-name>
        <url-pattern>/Hello</url-pattern>
    </servlet-mapping>

</web-app>
```

（4）测试 Servlet。启动 Tomcat，在浏览器中输入http://127.0.0.1:8080/HelloWorl/Hello，如果一切正常的话，将看到如图 5-3 所示的界面。

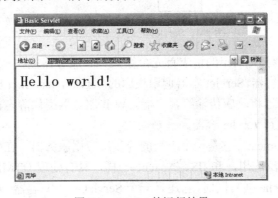

图 5-3  Servlet 的运行结果

## 5.3  Servlet 的体系结构

### 5.3.1  一个基本 Servlet 程序的组成

以清单 5-5 中的 HelloWorld 为例，分析一个基本的 Servlet 程序的组成部分。首先必须导入 javax.servlet.*和 javax.servlet.http.*，其中，javax.servlet.* 存放与 HTTP 协议无关的一般性 Servlet 类；javax.servlet.http.* 增加了与 HTTP 协议有关的功能。

所有Servlet都必须实现javax.servlet.Servlet接口,但是通常都会从javax.servlet.GenericServlet 或 javax.servlet.http.HttpServlet 择一来实现。如果编写的 Servlet 代码和 HTTP 协议无关，那么必须继承 GenericServlet 类；若有关，就必须继承 HttpServlet 类。HelloWorld 例子中继承的是 HttpServlet 类。

javax.servlet.* 里面的 ServletRequest 和 ServletResponse 接口提供存取一般的请求和响应；而 javax.servlet.http.* 里面的 HttpServletRequest 和 HttpServletResponse 接口，则提供 HTTP 请求及响应的存取服务。HelloWorld 示例代码中用到的是 HttpServletRequest 和 HttpServletResponse。

代码中利用 HttpServletResponse 接口的 setContentType()方法来设定内容类型, HelloWorld

示例中要显示为 HTML 网页类型，因此，内容类型设为 text/html，这是 HTML 网页的标准 MIME 类型值。之后，用 getWriter()方法返回 PrintWriter 类型的 out 对象，它与 PrintStream 类似，但是它能够对 Java 的 Unicode 字符进行编码转换。最后，利用 out 对象把"Hello world!"的字符串显示在网页上。

### 5.3.2 Servlet 应用程序体系结构

Servlet 容器将 Servlet 动态地加载到服务器上。HTTP Servlet 使用 HTTP 请求和 HTTP 响应标题与客户端进行交互。因此 Servlet 容器支持请求和相应所用的 HTTP 协议。Servlet 应用程序体系结构如图 5-4 所示。

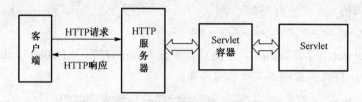

图 5-4  Servlet 应用程序体系结构

图 5-4 说明客户端对 Servlet 的请求首先会被 HTTP 服务器接收，HTTP 服务器将客户的 HTTP 请求提交 Servlet 容器，Servlet 容器调用相应的 Servlet，Servlet 作出的响应传递到 Servlet 容器，并由 HTTP 服务器将响应传输给客户端。Web 服务器提供静态内容并将所有客户端对 Servlet 作出的请求传递到 Servlet 容器。

我们已经学习过 Tomcat，它是一个小型的轻量级应用服务器，在中小型系统和并发用户不是很多的情况下被广泛应用，和 IIS、Apache 一样，具有处理 HTML 的功能（但处理静态 HTML 的能力不如 Apache 强）。同时，它还是一个 Servlet 和 JSP 容器，开发和调试 JSP、Servlet 的首选。对于图 5-4，Tomcat 就是 HTTP 服务器和 Servlet 容器两个部分。

### 5.3.3 Servlet 层次结构

Servlet 是实现 javax.servlet.Servlet 接口的对象。大多数 Servlet 通过从 GenericServlet 或 HttpServlet 类进行扩展来实现。Servlet API 包含于两个包中，即 javax.servlet 和 javax.servlet.http。各接口及其功能分别如表 5-1、表 5-2 所示。

表 5-1  javax.servlet

| 接口 | 功能 | 类 | 功能 |
| --- | --- | --- | --- |
| ServletConfig | 定义在 Servlet 初始化的过程中由 Servlet 容器传递给 Servlet 的配置信息对象 | ServletInputStream | 定义名为 readLine()的方法，从客户端读取二进制数据 |
| ServletContext | 定义 Servlet 使用的方法以获取其容器的信息 | ServletOutputStream | 向客户端发送二进制数据 |
| ServletRequest | 定义一个对象封装客户向 Servlet 的请求信息 | GenericServlet | 抽象类，定义一个通用的、独立于底层协议的 Servlet |
| ServletResponse | 定义一个对象辅助 Servlet 将请求的响应信息发送给客户端 | | |
| Servlet | 定义所有 Servlet 必须实现的方法 | | |

表 5-2  javax.servlet.http

| 接 口 | 功 能 | 类 | 功 能 |
| --- | --- | --- | --- |
| HttpSession | 用于标识客户端并存储有关客户端的信息 | HttpServlet | 扩展了 GenericServlet 的抽象类，用于扩展创建 Http Servlet |
| HttpSessionAttributeListener | 这个侦听接口用于获取会话的属性列表的改变的通知 | Cookie | 创建一个 Cookie，用于存储 Servlet 发送给客户端的信息 |
| HttpServletRequest | 扩展 ServletRequest 接口，为 HTTP Servlet 提供 HTTP 请求信息 | | |
| HttpServletResponse | 扩展 ServletResponse 接口，提供 HTTP 特定的发送响应的功能 | | |

## 5.4  Servlet 的生命周期

### 5.4.1  Servlet 的生命周期

Web 容器可以为到来的每一个请求创建一个 Servlet 实例，但是这种做法的效率不高。如今服务器每秒可以处理成千上万的请求，但创建一个 Servlet 实例要花费很多时间，而且会占用服务器上的大量内存。如果采用这种方式来实例化 Servlet，就会很快耗尽服务器的内存和处理资源。

为了更高效地处理到来的请求，容器必须对 Servlet 实例的创建进行优化。尽可能少地创建 Servlet 的实例，并重用这个实例来处理所有到来的请求。对所创建的实例进行适当的管理，服务器内存不太够时，要撤销较老的未用实例。

因此每一个 Servlet 都有一个生命期。它定义了一个 Servlet 如何被加载、初始化，以及它怎样接收请求、响应请求、提供服务，它准确地描述了在这些重要的生命事件期间 Servlet 和容器用何种方式进行交互。

在程序代码中，Servlet 生命周期由接口 javax.sevlet.Servlet 定义。所有的 Java Servlet 必须直接或间接地实现 javax.servlet.Servlet 接口，这样才能在 Servlet Engine 上运行。Servlet Engine 提供 network Service，响应 MIME request，运行 Servlet Container。javax.servlet.Servlet 接口定义了一些方法，在 Servlet 的生命周期中，这些方法会在特定时间按照一定的顺序被调用。Servlet 的生命周期如图 5-5 所示。

Servlet 运行在 Servlet 容器中，其生命周期由容器来管理。Servlet 的生命周期通过 javax.servlet.Servlet 接口中的 init()、service()和 destroy()方法来表示。

Servlet 的生命周期包含了下面 4 个阶段：

1. 加载和实例化

Servlet 容器负责加载和实例化 Servlet。当 Servlet 容器启动时，或者在容器检测到需要这个 Servlet 来响应第一个请求时，创建 Servlet 实例。当 Servlet 容器启动后，它必须要知道所需的 Servlet 类在什么位置，Servlet 容器可以从本地文件系统、远程文件系统或者其他的网络服务中通过类加载器加载 Servlet 类，成功加载后，容器创建 Servlet 的实例。因为容器是通过 Java 的反射 API 来创建 Servlet 实例，调用的是 Servlet 的默认构造方法（即不带参数的构造方法），所以我们在编写 Servlet 类的时候，不应该提供带参数的构造方法。

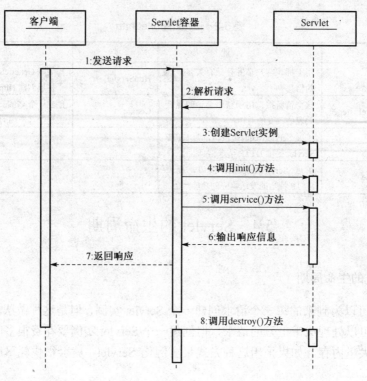

图 5-5 Servlet 的生命周期

### 2. 初始化

在 Servlet 实例化之后，容器将调用 Servlet 的 init()方法初始化这个对象。初始化的目的是为了让 Servlet 对象在处理客户端请求前完成一些初始化的工作，如建立数据库的连接，获取配置信息等。对于每一个 Servlet 实例，init()方法只被调用一次。在初始化期间，Servlet 实例可以使用容器为它准备的 ServletConfig 对象从 Web 应用程序的配置信息（在 web.xml 中配置）中获取初始化的参数信息。这样 Servlet 的实例就可以把与容器相关的配置数据保存起来供以后使用，在初始化期间，如果发生错误，Servlet 实例可以抛出 ServletException 异常，一旦抛出该异常，Servlet 就不再执行，而随后对它的调用会导致容器对它重新载入并再次运行此方法。

### 3. 请求处理

Servlet 容器调用 Servlet 的 service()方法对请求进行处理。要注意的是，在 service()方法调用之前，init()方法必须成功执行。在 service()方法中，通过 ServletRequest 对象得到客户端的相关信息和请求信息，在对请求进行处理后，调用 ServletResponse 对象的方法设置响应信息。对于 HttpServlet 类，该方法作为 HTTP 请求的分发器，这个方法在任何时候都不能被重载。当请求到来时，service()方法决定请求的类型（GET、POST、HEAD、OPTIONS、DELETE、PUT、TRACE），并把请求分发给相应的处理方法（doGet()、doPost()、doHead()、doOptions()、doDelete()、doPut()、doTrace()）每个 do 方法具有和第一个 service()相同的形式。我们常用的就是 doGet()和 doPost()方法，为了响应特定类型的 HTTP 请求，我们必须重载相应的 do 方法。如果 Servlet 收到一个 HTTP 请求而你没有重载相应的 do 方法，它就返回一个说明此方法对本资源不可用的标准 HTTP 错误。

4. 服务终止

当容器检测到一个 Servlet 实例应该从服务中被移除的时候,容器就会调用实例的 destroy() 方法,以便让该实例可以释放它所使用的资源,保存数据到持久存储设备中。当需要释放内存或者容器关闭时,容器就会调用 Servlet 实例的 destroy() 方法。在 destroy() 方法调用之后,容器会释放这个 Servlet 实例,该实例随后会被 Java 的垃圾收集器所回收。如果再次需要这个 Servlet 处理请求,Servlet 容器会创建一个新的 Servlet 实例。

## 5.4.2 Servlet 的基本方法

javax.servlet.Servlet 接口是容器(应用服务器)和 Servlet 之间的合约。所有 Servlet2.4(或 2.3)规范的容器都要使用这个接口来访问 Servlet 提供的特性。在该接口中定义了一些方法完成 Servlet 的加载、初始化、运行和销毁,其提供的基本方法如表 5-3 所示。

表 5-3　javax.servlet.Servlet 接口中的方法

| 方法 | 作用 |
| --- | --- |
| init() | 容器调用该方法来初始化 Servlet,所有的初始化参数都在这里处理 |
| destory() | 容器调用该方法来指示 Servlet,它的使命已经结束 |
| getServletInfo() | 容器或者工具使用该方法来得到有关 Servlet 的信息。返回值是一个字符串,其中可能包含开发商名、版本声明等信息 |
| getServletConfig() | 容器使用该方法来得到与这个 Servlet 实例相关联的 javax.sevlet.ServletConfig 对象。容器在调用 init() 方法时,这个对象会传递到 Servlet |
| service() | 这是最关键的 Servlet 方法。容器调用这个方法向 Servlet 传递一个请求来进行处理,Servlet 必须处理这个请求,并提供一个响应 |

但是在实际的应用当中,大多数的 Servlet 并没有直接实现 javax.servlet.Servlet 接口,相反,它只是扩展了一个辅助类,而这个辅助类实现了该接口。用的最多的辅助类是 javax.servlet.HttpServlet,它实现了所需要的所有的方法。程序员要做的仅仅是覆盖要修改的方法。javax.servlet.HttpServlet 该抽象类会透明地对 HTTP 协议细节进行解码,使程序员能够把重点放在 Servlet 的核心功能上。HttpServlet 提供 doGet()、doPost()、doPut()、doDelete()、init()、serviece()、destory()、getServletInfo() 等方法。作为 javax.servlet.HttpServlet 抽象类的任何一个子类,可以覆盖其任何一个方法,来完成自己的某些特定功能。

1. init() 方法

在 Servlet 的生命期中,仅执行一次 init() 方法。它是在服务器装入 Servlet 时执行的,可以配置服务器,以在启动服务器或客户机首次访问 Servlet 时装入 Servlet。无论有多少客户机访问 Servlet,都不会重复执行 init()。

缺省的 init() 方法通常是符合要求的,但也可以用定制 init() 方法来覆盖它,典型的是管理服务器端资源。例如,可能编写一个定制 init() 来只用于装入一次 GIF 图像,改进 Servlet 返回 GIF 图像和含有多个客户机请求的性能。另一个示例是初始化数据库连接。缺省的 init() 方法设置了 Servlet 的初始化参数,并用它的 ServletConfig 对象参数来启动配置,因此所有覆盖 init() 方法的 Servlet 应调用 super.init() 以确保仍然执行这些任务。在调用 service() 方法之前,应确保已完成了 init() 方法。

2. service()方法

service()方法是Servlet的核心。每当一个客户请求一个HttpServlet对象,该对象的service()方法就要被调用,而且传递给这个方法一个请求（ServletRequest）对象和一个响应（ServletResponse）对象作为参数。

在HttpServlet中已存在service()方法。缺省的服务功能是调用与HTTP请求的方法相应的do功能。例如,如果HTTP请求方法为GET,则缺省情况下就调用doGet()。Servlet应该为Servlet支持的HTTP方法覆盖do功能。因为HttpServlet.service()方法会检查请求方法是否调用了适当的处理方法,不必要覆盖service()方法。只需覆盖相应的do方法就可以了。

3. destroy()方法

destroy()方法仅执行一次,即在服务器停止且卸装Servlet时执行该方法。典型地,将Servlet作为服务器进程的一部分来关闭。缺省的destroy()方法通常是符合要求的,但也可以覆盖它,典型的是管理服务器端资源。例如,如果Servlet在运行时要累计统计数据,则可以编写一个destroy()方法,该方法用于在未装入Servlet时将统计数字保存在文件中。另一个示例是关闭数据库连接。

当服务器卸载Servlet时,将在所有service()方法调用完成后,或在指定的时间间隔过后调用destroy()方法。一个Servlet在运行service()方法时可能会产生其他的线程,因此请确认在调用destroy()方法时,这些线程已终止或完成。

4. GetServletConfig()方法

GetServletConfig()方法返回一个ServletConfig对象,该对象用来返回初始化参数和ServletContext。ServletContext接口提供有关Servlet的环境信息。

5. GetServletInfo()方法

GetServletInfo()方法是一个可选的方法,它提供有关Servlet的信息,如作者、版本、版权。

6. doGet()方法

处理通过HTTP GET动作发送数据的到来的请求。当一个客户通过HTML表单发出一个HTTP GET请求或直接请求一个URL时,doGet()方法被调用。与GET请求相关的参数添加到URL的后面,并与这个请求一起发送。当不会修改服务器端的数据时,应该使用doGet()方法。

7. doPost()方法

处理通过HTTP POST动作发送数据的到来的请求。当一个客户通过HTML表单发出一个HTTP POST请求时,doPost()方法被调用。与POST请求相关的参数作为一个单独的HTTP请求从浏览器发送到服务器。当需要修改服务器端的数据时,应该使用doPost()方法。

8. doPut()方法

处理通过HTTP PUT动作发送数据的到来的请求（这个动作很少使用）。

9. doDelete()

处理通过HTTP DELETE动作删除服务器内容的到来的请求（这个动作很少使用）。

## 5.5 JSP 和 Servlet 的交互

Servlet 功能强大，能够完成各种任务，比如：读取用户通过 HTML 表单提交的数据、读取由浏览器发送的隐式数据、计算响应结果、向客户发送各种格式的数据(包括 HTML 文档)、发送隐式的 HTTP 响应数据等。因此，Servlet 与 JSP 之间的交互为开发 Web 服务提供了优秀的解决方案。

### 5.5.1 通过表单向 Servlet 提交数据

在 Servlet 中，可以读取用户通过 HTML 表单提交的数据。HttpServletRequest 提供了 4 个方法获取表单数据，如表 5-4 所示。

表 5-4 HttpServletRequest 获得参数名和参数值的方法

| 方法 | 说明 |
| --- | --- |
| getParameter(String key) | 返回一个字符串，获得 name 和 key 一样的表单控件的数据，如果有重复的 name，则返回第一个的值 |
| getParameterValues(String key) | 返回一个字符串数组，获得 name 和 key 一样的表单控件的数据，但相同 name 的控件会有多个，如同名的多个 checkbox 等 |
| getParameterMap() | 返回一个包含所有参数的 Map，为 key-String[]模式，即，key 是表单控件的 name，同时，为了防止有重复 name 的控件存在，每个 name 对应的值是一个字符串数组 |
| getParameterNames() | 返回一个枚举类型值，返回所有表单中所有表单控件的 name |

读取请求（表单）的参数时，只需调用 HttpServletRequest 的 getParameter 方法，提供大小写敏感的参数名作为方法的参数。只要所提供的参数名与 HTML 源代码中出现的参数名完全相同，就可以得到与终端用户的输入完全一致的结果。不论数据是以 GET 方式发送，还是以 POST 方式发送，都可以用完全相同的方式使用 getParameter。Servlet 知道客户使用的是哪种请求，自动使用恰当的方法读取数据。如果参数存在但没有相应的值（即用户在提交表单时没有填写对应的文本字段），则返回空的字符串；如果没有这样的参数，则返回 null。

如果同一参数名有可能在表单数据中多次出现，则应该调用 getParameterValues 方法，它返回字符串数组，而不是调用 getParameter，getParameter 方法仅返回对应参数首次出现的值。对于不存在的参数名，getParameterValues 方法返回 null，如果参数只有单一的值，则返回只有一个元素的数组。比如要获得多选框列表（即设置了 MULTIPLE 属性的 HTML SELECT 元素）中选定的元素的值，必须使用 getParameterValues 方法。

下面举例说明这 4 个方法的用法。例子中通过表单向 Servlet 传递数据的步骤及方法。该例中 Servlet 从表单数据中获取相应的 3 个参数，通过 Servlet 显示在 HTML 页面上。

（1）新建 Web 应用，例如 transfer，放在 tomcat 的 webapps 目录下。
（2）新建表单数据页面 sendparam.html，放在/transfer 目录下，代码如清单 5-7 所示。

清单 5-7 sendparam.html 页面

```
<html>
<head>
<meta http-equiv="Content-Type" content="text/html; charset=UTF-8">
<title>测试 HttpRequest 接收参数方法_发送参数</title>
</head>
```

```html
<body>
<form action="GetParam" method="get">
发送的内容：<br>
输入框A(name="txt"):<input type="text" name="text"/>
<br>
输入框B(name="txt"):<input type="text" name="text"/>
<br><br>
<input type="checkbox" name="checkbox" value="a">
选项1(name="chk" value="a")
<br>
<input type="checkbox" name="checkbox" value="b">
选项2(name="chk" value="b")
<br><br>
<select name="select">
  <option value="1">内容1_值是1</option>
  <option value="2">内容2_值是2</option>
</select>
<br><br>
<input type="submit" value="传送"/>
</form>
</body>
</html>
```

在 sendparam.html 中，定义了两个输入框，name 都设定为 text，定义了两个多选框，name 都设定为 checkbox，最后定义了一个下拉选择框，name 设定为 select。

（3）编写 Servlet 类，例如 GetParamServlet.java，里边分别用到了上面提到的 4 个方法，程序代码如清单 5-8 所示。该 Servlet 的作用是获取表单传过来的参数值，并把它显示在 html 页面中，相关说明见程序注释。

（4）编译 Servlet 类 GetParamServlet.java（用 javac 编译时需要将 Servlet-api.jar 包加入系统的环境变量中），把生成的 class 文件 GetParamServlet.class 放置于 Web 应用 /transfer/WEB-INF/classes 目录下。

清单 5-8　GetParamServlet 类

```java
/*
 * GetParamServlet.java
 * 功能：Servlet 获得参数
 */
import java.io.IOException;
import java.io.PrintWriter;
import java.util.Enumeration;
import java.util.Iterator;
import java.util.Map;

import javax.servlet.http.HttpServlet;
import javax.servlet.http.HttpServletRequest;
import javax.servlet.http.HttpServletResponse;
```

```java
/**
 * Servlet 获得参数的方法演示
 */
public class GetParamServlet extends HttpServlet {

    /** SerialVersionUID */
    private static final long serialVersionUID = 1711689663622072980L;

    /**
     * 处理 Get 请求
     * @param req Request
     * @param resp Response
     * @throws IOException IO 异常
     */
    @Override
    protected void doGet(HttpServletRequest req,
            HttpServletResponse resp) throws IOException {

        // 设置输出的格式
        resp.setContentType("text/html;charset=UTF-8");
        PrintWriter out = resp.getWriter();
        out.println("<html>");
        out.println("<head>");
        out.println("<title>HttpRequest 获得参数的方法</title>");
        out.println("</head>");
        out.println("<body>");
        // 设置接收参数所用的编码
        req.setCharacterEncoding("UTF-8");

        // 通过 getParameter() 获得的参数
        out.println("getParameter() 获得的参数<br>");
        String value = null;
        // 获得 name="txt"的表单控件的值
        value = req.getParameter("text");
        out.println("输入框内输入的值:" + value);
        out.print("<br>");
        // 获得 name="chk"的表单控件的值
        value = req.getParameter("checkbox");
        out.println("多选框选中的值:" + value);
        out.print("<br>");
        // 获得 name="sel"的表单控件的值
        value = req.getParameter("select");
        out.println("下拉框选中的值:" + value);
        out.println("<br>");
        out.println("====================================<br>");

        // 通过 getParameterValues() 获得的参数
        out.println("getParameterValues() 获得的参数<br>");
```

```java
        // 获得 name="txt"的表单控件的值
        String[] paramValue = null;
        paramValue = req.getParameterValues("text");
        if (null != paramValue) {
            out.print("输入框内输入的值:");
            for (int i = 0; i < paramValue.length; i++) {
                out.print(paramValue[i]);
                out.print("  ");
            }
        }
        out.println("<br>");
        // 获得 name="chk"的表单控件的值
        paramValue = req.getParameterValues("checkbox");
        if (null != paramValue) {
            out.print("多选框选中的值:");
            for (int i = 0; i < paramValue.length; i++) {
                out.print(paramValue[i]);
                out.print("  ");
            }
        }
        out.println("<br>");
        // 获得 name="sel"的表单控件的值
        paramValue = req.getParameterValues("select");
        if (null != paramValue) {
            out.print("下拉框选中的值:");
            for (int i = 0; i < paramValue.length; i++) {
                out.print(paramValue[i]);
                out.print("  ");
            }
        }
        out.println("<br>");
        out.println("======================================<br>");

        // 通过 getParameterMap()获得的参数
        out.println("getParameterMap()获得的参数<br>");
        // 获得参数的 Map
        Map paramMap = req.getParameterMap();
        Object o = null;
        String[] val = null;
        // 循环 Map 的 key
        for (Iterator it = paramMap.keySet()
                .iterator(); it.hasNext();) {
            o = it.next();
            out.print(o);
            out.print(":");
            // 获得 key 对应的 value
            val = (String[]) paramMap.get(o);
            if (null != val) {
```

```java
            for (int j = 0; j < val.length; j++) {
                out.print(val[j]);
                out.print("  ");
            }
        }
    }
    out.println("<br>");
    out.println("=====================================<br>");
    // 通过getParameterNames()获得参数的名称
    out.println("getParameterNames()获得参数的名称<br>");
    // 获得参数名称枚举
    Enumeration en = req.getParameterNames();
    for (;en.hasMoreElements();) {
        o = en.nextElement();
        out.print(o);
        out.print("  ");
    }
    out.println("<br>");
    out.println("=====================================<br>");
    out.println("</body>");
    out.println("</html>");
    }
}
```

（5）部署 Web.xml。修改/transfer/WEB-INF 目录下的部署描述符 web.xml，添加 Servlet 映射，在其中加入清单 5-9 的片段。

清单 5-9　web.xml 片段

```xml
<servlet>
    <servlet-name>GetParam</servlet-name>
    <servlet-class>GetParamServlet </servlet-class>
</servlet>
<servlet-mapping>
    <servlet-name>GetParam</servlet-name>
    <url-pattern>/GetParam</url-pattern>
</servlet-mapping>
```

在 Servlet 里，每个方法都会把 sendparam.html 中所有的表单控件的值取出来并表示。对于除 getParameter(String key) 以外的其他 3 个方法 getParameterValues(String key)、getParameterNames()和 getParameterMap()而言，只要在表单中输入的值都会被获取。而对于 getParameter(String key)而言，在有重复 name 控件的情况下，它只会获得第一个控件的值。

（6）测试 Servlert。启动 Tomcat，在浏览器中输入http://localhost:8080/transfer/sendparam.html，页面如图 5-6 所示，输入图中所示的数据。

输出结果如图 5-7 所示。

示例中通过 getParameter()获得参数的时候，对于输入框，由于第一个输入框没有输入任何值，所以，取到得值就是空值；而对于多选框，由于只会取得第一个控件的值，所以只是输出了第一个多选框的值：选项值1。通过 getParameterValues()和 getParameterMap()获得参数时，由于值都放在字符串数组中，所以，只要是输入过的值，都可以被获取到并显示。

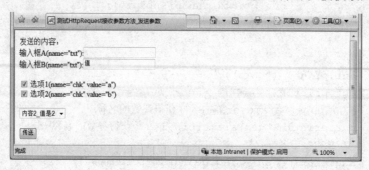

图 5-6  在表单中输入数据

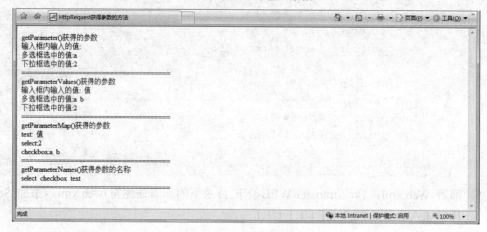

图 5-7  获得参数输出结果

### 5.5.2 从 Servlet 到 JSP 的信息传递

通过上节了解了在 Servlet 中获取 JSP 表单信息的方法。当然 Servlet 中的信息一样可以通过某种方式传递到 JSP 中，从而达到 Servlet 与 JSP 之间的数据共享。

从 Servlet 到 JSP 间信息传递的过程，实际上就是在 Servlet 中把信息存储在某一范围的对象属性中，然后在 JSP 中通过对应的方法把该范围内的对象属性取出来的过程。在 Servlet 中，提供存储信息的类或接口有 HttpServletRequest、HttpSession、ServletContext 等，它们分别提供了 4 个读取或设置共享数据的方法。

（1）setAttribute(String name，Object object)：把一个对象和一个属性名绑定，将这个对象存储在 HttpServletRequest 或 HttpSession 或 ServletContex 中。如果指定的名称已经被使用过，这个方法会删除旧对象绑定为新对象。

（2）getAttribute(String name)：根据指定的属性名返回所绑定的对象，如果名称不存在，返回 null。

（3）removeAttribute(String name)：根据给定的属性名从 HttpServletRequest 或 HttpSession 或 ServletContex 中删除指定名称的对象。

（4）getAttributeNames()：返回 Enumeration 对象，它包含了存储在相应对象中的所有属性名。

从 Servlet 到 JSP 的信息传递，只需要在 Servlet 中把想要传递的信息通过 setAttribute(String name, Object object)方法存入相应范围的对象属性中，在 JSP 中，通过相应的方法取出对象值就可以了。在 Servlet 中，存入 HttpServletRequest 中的对象值，在 JSP 中，通过 JSP 的内置对象 request 的 getAttribute(String name)方法就可以把相应的对象值取出来。存入 HttpSession 中的对象值，在 JSP 中，通过 JSP 的内置对象 session 的 getAttribute(String name)方法就可以把相应的对象值取出来。存入 servletContext 中的对象值，在 JSP 中，通过 JSP 的内置对象 application 的 getAttribute(String name)方法就可以把相应的对象值取出来。

为了更好地使用它们，需要了解一下它们的作用范围。

HttpServletRequest 对象的作用范围最小。时间上：只是本身请求和应答完成就失效，当然转发是把当前的 request 对象取出来传给另一资源，其实本身的 request 对象还是只生存到本次请求结束。空间上：只能发送请求的客户端有效。

HttpSession 对象的时间作用范围比 HttpServletRequest 大，空间作用范围相同。它的作用范围是一次连接至到客户端关闭。

Servlet 容器在启动时会加载 Web 应用，并为每个 Web 应用创建唯一的 ServletContext 对象。可以把 ServletContext 看成是一个 Web 应用的服务器端组件的共享内存。在 ServletContext 中可以存放共享数据。它对任何 Servlet，任何人在任何时间都有效，是真正的全局对象。

下面举例演示从 Servlet 向 JSP 的信息传递的步骤及方法。该例在 Servlet 中为不同范围的对象属性存入相应的信息，在 JSP 中获取这些信息并把它们的值显示在 HTML 页面上。

（1）新建 Web 应用，例如 test，放在 tomcat 的 webapps 目录下。

（2）编写 Servlet 类，例如 store.java，程序代码如清单 5-10 所示。该 Servlet 的作用是分别在 HttpServletRequest、HttpSession、ServletContext 对象范围内存入相应的信息，黑体部分为具体的存入对象的具体方法

清单 5-10  store.java

```
import javax.servlet.*;
import javax.servlet.http.*;
import java.io.*;

public class store extends HttpServlet{
  private static final String CONTENT_TYPE = "text/html; charset=GBK";
  public void service(HttpServletRequest request, HttpServletResponse
                 response) throws
     ServletException, IOException {
       response.setContentType(CONTENT_TYPE);
       request.setCharacterEncoding("GBK");
       String target = "/show.jsp";

       request.setAttribute("requestTest","这是存放在request范围内的测试");
       HttpSession session = request.getSession();
       session.setAttribute("sessionTest","这是存放在session范围内的测试");
       ServletContext context=getServletContext();
       context.setAttribute("contextTest","这是存放在context范围内的测试");
```

```
            RequestDispatcher dispatcher=context.getRequestDispatcher(target);
            dispatcher.forward(request,response);
        }
    }
```

（3）编译 Servlet 类 store.java（用 javac 编译时需要将 Servlet-api.jar 包加入系统的环境变量中），把生成的 class 文件 store.class 放置于 Web 应用/transfer/WEB-INF/classes 目录下。

（4）部署 Web.xml。修改/test/WEB-INF 目录下的部署描述符 Web.xml，添加 Servlet 映射，如清单 5-11 所示，黑体部分为注册添加的代码。0

清单 5-11   web.xml

```xml
<?xml version="1.0" encoding="ISO-8859-1"?>

<web-app xmlns="http://java.sun.com/xml/ns/javaee"
   xmlns:xsi="http://www.w3.org/2001/XMLSchema-instance"
   xsi:schemaLocation="http://java.sun.com/xml/ns/javaee
                       http://java.sun.com/xml/ns/javaee/web-app_2_5.xsd"
   version="2.5">
  <display-name>Welcome to Tomcat</display-name>
  <description>
     Welcome to Tomcat
  </description>

<servlet>
    <servlet-name>store</servlet-name>
    <servlet-class>store</servlet-class>
</servlet>
<servlet-mapping>
    <servlet-name> store </servlet-name>
    <url-pattern>/store</url-pattern>
</servlet-mapping>

</web-app>
```

（5）新建数据显示页面 show.jsp，放在/test 目录下，其代码如清单 5-12 所示。该页面为从 Servlet 中获取相应信息的显示页面。

清单 5-12   show.jsp

```jsp
<%@ page contentType="text/html; charset=gb2312" language="java"%>
<html>
<body>
<center>从 servlet 中取值结果如下:
<br>
<br>
request 中的值为:
```

```
<%=request.getAttribute("requestTest")%>
<br>
<br>
session 中的值为：
<%=session.getAttribute("sessionTest")%>
<br>
<br>
context 中的值为：
<%=application.getAttribute("contextTest")%>
</body>
</html>
```

（6）测试 Servlet。启动 Tomcat，在浏览器中输入http://localhost:8080/test/store，如果一切正常的话，将出现如图 5-8 所示的页面。

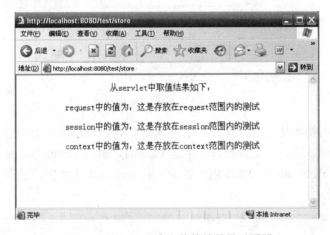

图 5-8  从 Servlet 中取值的结果显示页面

## 5.6  Servlet 的高级应用

Servlet 的高级应用可以为程序编写提供很大的便利，包括 Servlet 初始化参数的使用、Servlet 过滤器 Filter 和 Servlet 监听器 Listener。本节重点学习 Servlet 初始化参数的配置方法和 Servlet 中读取参数的方法，Filter 的工作原理，在 web.xm 中的配置方法以及开发方法，Listener 的工作原理，在 web.xml 中的配置方法和开发方法。

### 5.6.1  Servlet 的初始化参数

在 Web 应用程序的配置文件（web.xml）中可以定义一些初始化参数，然后在 Servlet 中获得这些初始的参数。这样做的好处是使得 Servlet 的一些参数是可配置的，要修改这些参数，不需要修改源程序和重新编译，只需要修改 web.xml 即可，如数据库连接的 URL 等。

Eclipse 可以帮助开发者管理 web.xml 文件。有两种方式可以定义这些初始化参数，这两种方式都是修改 web.xml 文件中的内容。第一种方式是在 Servlet 创建时设置，第二种方式是直接编辑 web.xml 文件。

1. 在 Servlet 创建时定义参数

在创建 Servlet 时，输入 Servlet 的类名和包以后，单击 Next 按钮，会出现如图 5-9 所示的界面。

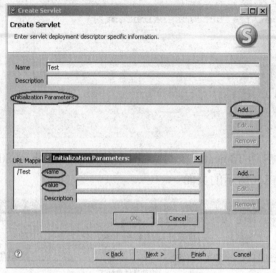

图 5-9  设置 Servlet 初始参数

中间的框中列出初始化参数。单击右边的 Add 按钮可以添加一个参数，每个参数至少包括名字和值。我们可以输入一个参数，名字是 Param1，值是 Param1Value。这样，在 Servlet 创建后，在项目的 WebContent\WEB-INF 目录下的 web.xml 文件中，就保存了这些参数的设置，如图 5-10 所示。

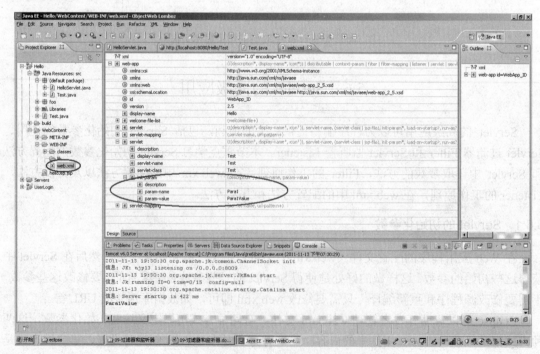

图 5-10  查看参数设置

打开 web.xml 后，可以管理 Servlet 的初始化参数。在 servlet 上右击，选 Add Child→init-param，可以创建一个新的参数。在参数的 param-name 右边的框中可以修改参数的名字；在参数的 param-value 右边的框中可以修改参数的值。在 init-param 上右击，选 Remove 可以删除该参数。

再创建一个参数，名字叫 Param2，值是 Param2Value。

2. 在 Servlet 中读取参数

要读取 web.xml 中的初始化参数，可以使用 Servlet 的 getInitParameter 方法，其用法类似于 request 对象的 getParameter 方法，其输入参数为参数名称，返回值为对应参数的值。

如：在 Servlet 的 doGet 方法中可以执行如下代码：

```
String para1 = this.getInitParameter("Param1");
System.out.println(para1);
```

则该 Servlet 执行时，将在 Tomcat 控制台输出 Param1Value 字符串。

3. Servlet 的配置

我们可以看出，web.xml 中不仅保存了 Servlet 的参数，还保存了网站的所有设置，包括名称映射等。Eclipse 可以帮助我们管理这些内容，防止出现人为错误。开发时，web.xml 在 Eclipse 启动 Tomcat 时，作为配置使用。实际部署时，web.xml 会直接部署到相应目录下，由 Tomcat 使用。

了解了 Servlet 如何配置，我们就能知道 Servlet 作为一个 Java 类是如何被浏览器请求到的。下面的代码 5-13 给出了创建了 Servlet 后 web.xml 中增加的信息。

清单 5-13　web.xml 片段

```
<servlet>
  <description></description>
  <display-name>Test</display-name>
  <servlet-name>Test</servlet-name>
  <servlet-class>Test</servlet-class>
  <init-param>
    <description></description>
    <param-name>Para1</param-name>
    <param-value>Para1Value</param-value>
  </init-param>
  <init-param>
    <param-name>Para2</param-name>
    <param-value>Para2Value</param-value>
  </init-param>
</servlet>
<servlet-mapping>
  <servlet-name>Test</servlet-name>
  <url-pattern>/Test</url-pattern>
</servlet-mapping>
```

其中：servlet 节点配置了 Web 应用程序中的一个 Servlet，其中的 servlet-class 代表具体的 Servlet 类的完整的包路径；servlet-mapping 节点定义了 Servlet 的映射信息，即将一个 URL 或某种 URL 模式映射到一个指定的 Servlet。每个 Servlet 节点都对应一个 Servlet-mapping 节点，且两个节点之间通过 servlet-name 建立对应关系。

在本例中，我们在浏览器中输入 http://localhost:8080/Hello/Test，Tomcat 收到请求后，发现该请求与 servlet-mapping 中的 Test 的 url-pattern 一致，于是 Tomcat 根据 servlet 节点中的配置，调用 name 为 Testt 的 Servlet，并根据 servlet-class 的配置调用相应的 servlet 类。

url-pattern 的设置中，除了上述的精确配置方法外，还有目录匹配和扩展名匹配两种方式，其具体格式为：

目录格式匹配：<url-pattern>/目录/*</url-pattern>

扩展名匹配：<url-pattern>*.扩展名</url-pattern>

Tomcat 在查找 Servlet 的映射时，遵守以下重要规则：

- 容器会首先查找精确匹配，如果找不到，再查找目录匹配，如果也找不到，就查找扩展名匹配。
- 如果一个请求匹配多个"目录匹配"，容器会选择最长的匹配。

4. 实际应用

实际应用中，比如我们要做一个使用 JDBC 连接数据库的 Servlet 程序，为了保证这个 Servlet 的通用性，可以把它的驱动程序、连接数据库的字符串配置到 web.xml 中。当 Servlet 执行操作时，可以通过读取 web.xml 的配置信息来获取数据库的连接信息。

### 5.6.2 过滤器

1. 过滤器基本原理

过滤器（Filter）是 Servlet 2.3 规范中新增加的，是实现 Filter 接口的一个类，该类截取用户从客户端提交的请求，在还没有到达需要访问的资源时运行。它操纵来自客户端的请求，在资源还没有被发送到客户端前截取响应，并处理这些还没有发送到客户端的响应，原理如图 5-11 所示。

需要注意的是，Filter 并不是一个 servlet，它不能产生一个 response，它能够在一个请求（request）到达服务器之前预处理该请求，也可以在离开服务器时处理响应（response），如图 5-12 所示。当有多个过滤器时，多个过滤器之间构成过滤器链（FilterChain）。

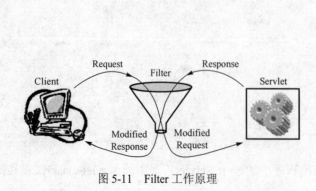

图 5-11　Filter 工作原理

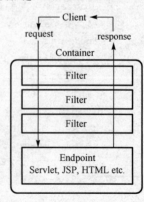

图 5-12　过滤器链

## 2. 创建一个简单的过滤器

要创建过滤器，首先创建一个实现 Filter 接口的类，然后将该类配置到 web.xml 中，以保证过滤器在 web 应用程序启动时既可加载。以下给出创建一个简单过滤器的步骤：

（1）创建一个实现 Filter 接口的类：SimpleFilter。在 FirstWebApp 项目的 src 目录上右击，选择 New→Class，在 package 输入框中输入包名，如 test.Filter，在 Name 输入框中输入类名 SimpleFilter，然后单击 Add 按钮，弹出如图 5-13 所示对话框。在输入框中输入 filter，在自动列出的接口列表中选择 Filter – javax.servlet … 接口，单击 OK 按钮，再单击 Finish 按钮，一个空的过滤器被创建出来。

在创建的 SimpleFilter 类中，共包含 3 个方法，分别是：init 方法、destroy 方法和 doFilter 方法。其中，init 方法和 destroy 方法分别执行过滤器的初始化和销毁操作，而过滤器的具体业务操作则由 doFilter 方法实现。

doFilter 方法共有 3 个参数，其类型分别是 ServletRequest、ServletResponse 和 FilterChain。由图 5-5 及 Filter 的基本原理，过滤器首先对用户请求进行过滤，然后将请求传递给服务器（Tomcat），服务器返回一个响应（Response）时，也要被过滤器过滤后才返回客户端。

图 5-13　接口选择对话框

因此可知，doFilter 中应首先对 request 进行处理，然后将请求继续传递给服务器（Tomcat），最后对 response 进行处理。

因此，我们在 doFilter 方法中添加如下代码：

```
System.out.println("Processing Request");  // 这部分写针对request对象处理的代码
arg2.doFilter(arg0, arg1);//将请求继续传递给下一个过滤器或服务器（Tomcat）
System.out.println("Processing Response");//对服务器返回的response对象进行处理
```

（2）在 web.xml 中配置 SimpleFilter 类。

要想让 Filter 起作用，还要将其配置到 web.xml 中。在 web-app 上右击，选 Add Child→context-param-login-config→filter 和 filter-mapping，在 web.xml 中创建出相应的项，然后再填写 Filter 的配置。

Filter 的配置与 Servlet 的配置类似，如清单 5-14 所示。

清单 5-14　Filter 在 web.xml 中的配置

```
<filter>
    <filter-name>SimpleFilter</filter-name>
    <filter-class>edu.sdfi.train.filters.SimpleFilter</filter-class>
</filter>
<filter-mapping>
    <filter-name>SimpleFilter</filter-name>
    <url-pattern>/*</url-pattern>
</filter-mapping>
```

（3）重新启动 Tomcat，在地址栏输入任何一个请求，然后查看 Tomcat 控制台的输出。正常运行后，其效果如图 5-14 所示。

图 5-14 过滤器执行后的控制台输出

让过滤器过滤不同的资源，可以有以下几种不同的方式：

（1）映射过滤应用程序中所有资源。

```
<filter>
    <filter-name>loggerfilter</filter-name>
    <filter-class>myfilter.LoggerFilter</filter-class>
</filter>
<filter-mapping>
    <filter-name>loggerfilter</filter-name>
    <url-pattern>/*</url-pattern>
</filter-mapping>
```

（2）过滤指定的类型文件资源

```
<filter>
    <filter-name>loggerfilter</filter-name>
    <filter-class>myfilter.LoggerFilter</filter-class>
</filter>
<filter-mapping>
    <filter-name>loggerfilter</filter-name>
    <url-pattern>*.html</url-pattern>
</filter-mapping>
```

其中<url-pattern>*.html</url- pattern>要过滤 JSP，那么就改*.html 为*.jsp，但是注意没有"/"斜杠。

（3）同时过滤多种类型资源。

```
<filter>
    <filter-name>loggerfilter</filter-name>
    <filter-class>myfilter.LoggerFilter</filter-class>
</filter>
<filter-mapping>
    <filter-name>loggerfilter</filter-name>
    <url-pattern>*.html</url-pattern>
</filter-mapping>
<filter-mapping>
    <filter-name>loggerfilter</filter-name>
    <url-pattern>*.jsp</url-pattern>
</filter-mapping>
```

（4）过滤指定的目录。

```
<filter>
    <filter-name>loggerfilter</filter-name>
    <filter-class>myfilter.LoggerFilter</filter-class>
</filter>
```

```xml
<filter-mapping>
    <filter-name>loggerfilter</filter-name>
    <url-pattern>/folder_name/*</url-pattern>
</filter-mapping>
```

(5) 过滤指定的 Servlet。

```xml
<filter>
    <filter-name>loggerfilter</filter-name>
    <filter-class>myfilter.LoggerFilter</filter-class>
</filter>
<filter-mapping>
    <filter-name>loggerfilter</filter-name>
    <servlet-name>loggerservlet</servlet-name>
</filter-mapping>
<servlet>
    <servlet-name>loggerservlet</servlet-name>
    <servlet-class>myfilter.LoggerServlet</servlet-class>
</servlet>
```

(6) 过滤指定文件。

```xml
<filter>
    <filter-name>loggerfilter</filter-name>
    <filter-class>myfilter.LoggerFilter</filter-class>
</filter>
<filter-mapping>
    <filter-name>loggerfilter</filter-name>
    <url-pattern>/simplefilter.html</url-pattern>
</filter-mapping>
```

以上都要注意是否有斜杠"/"。

3. 使用 Servlet 过滤器实现用户登录验证

在一个系统中，往往第一步就是让用户登录，根据用户读出权限，然后列出菜单供用户操作。用户登录后一般将其信息存储到 session 中，在其他的页面从 session 中读取用户信息，如果不存在，则表明用户并未登录，就跳转到登录页面要求用户登录。

如果不使用过滤器的话当然也可以实现，那就要在每一个页面添加验证信息，这样就很复杂，而且不利于管理。我们的宗旨是，只要是能够统一处理的，就一定要将这个功能作为公共模块提取出来。实现此功能的代码如清单 5-15 所示。

清单 5-15 用户登录验证过滤器

```java
package Filters;
import javax.servlet.FilterChain;
import javax.servlet.FilterConfig;
import javax.servlet.ServletRequest;
import javax.servlet.ServletResponse;
import javax.servlet.http.HttpServletRequest;
import javax.servlet.http.HttpServletResponse;
```

```java
import java.io.*;
import javax.servlet.*;
import javax.servlet.http.*;
public class LogOrNot implements javax.servlet.Filter {
private FilterConfig config;
private String logon_page;
private String home_page;
public void destroy() {
config = null;
}
public void init(FilterConfig filterconfig) throws ServletException {
// 从部署描述符中获取登录页面和首页的URI
config = filterconfig;
logon_page = filterconfig.getInitParameter("LOGON_URI");
home_page = filterconfig.getInitParameter("HOME_URI");
System.out.println(home_page);
if (null == logon_page || null == home_page) {
throw new ServletException("没有找到登录页面或主页");
}
}
public void doFilter(ServletRequest request, ServletResponse response,
FilterChain chain) {
HttpServletRequest req = (HttpServletRequest) request;
HttpServletResponse rpo = (HttpServletResponse) response;
javax.servlet.http.HttpSession session = req.getSession();
try {
req.setCharacterEncoding("utf-8");
} catch (Exception e1) {
e1.printStackTrace();
}
String userId = (String) session.getAttribute("UserId");
String request_uri = req.getRequestURI().toUpperCase();// 得到用户请求的URI
String ctxPath = req.getContextPath();// 得到web应用程序的上下文路径
String uri = request_uri.substring(ctxPath.length());
    // 去除上下文路径，得到剩余部分的路径
try {
if (request_uri.indexOf("LOGIN.JSP") == -1 && request_uri.indexOf
    ("LOG.JSP") == -1 && userId == null)
{
rpo.sendRedirect(home_page+logon_page);
System.out.print(home_page+logon_page);
return;
}
else {
chain.doFilter(request, response);
}
```

```
        } catch (Exception e) {
            e.printStackTrace();
        }
    }
}
```

对上面的代码稍作解释：过滤器从配置文件中读出配置选项，一个是登录页面的 URL 地址，另外一个是 Web 应用的 URL，之所以要这样做，是因为如果程序中有 iframe 的话，登录页面会默认在 iframe 中打开，因此这里将使用绝对地址进行跳转；判断语句中要将 login.jsp 和 log.jsp 排除，因为这两个是处理登录的页面，若不排除将出现循环重定向；检查 session 中的 userid 选项，当然也可以设置其他的，关键看 session 中存的是什么了，若有，则提交给下一个过滤器，若不再有过滤器，则提交给处理页面，若未登录，则跳转到登录页面。

代码写好后，在 Web 应用的 web.xml 文件中进行配置，如清单 5-16 所示。

清单 5-16　用户登录验证过滤器配置

```xml
<filter>
    <filter-name>LogOrNot</filter-name>
    <filter-class>Filters.LogOrNot</filter-class>
    <init-param>
        <param-name>LOGON_URI</param-name>
        <param-value>log.jsp</param-value>
    </init-param>
    <init-param>
        <param-name>HOME_URI</param-name>
        <param-value>/model/</param-value>
    </init-param>
</filter>
<filter-mapping>
    <filter-name>LogOrNot</filter-name>
    <url-pattern>*.jsp</url-pattern>
</filter-mapping>
```

其中两个配置参数就对应了 Java 文件中使用的参数 log.jsp 是登录的视图页面，/model/ 是当前 Web 应用的文件夹名称。mapping 里面定义了对*.jsp 进行过滤，如果程序中还有其他的页面，如.do 或者.action 等，那么可以继续添加<filter-mapping>这个选项，其中<url-pattern>就填写*.do 或者*.action 即可。

这样对所有的页面都可以进行过滤。当然还可以为应用配置其他的过滤器，Tomcat 容器会根据 web.xml 文件中的配置顺序将其设置称过滤器链挨个处理，处理到最后一个跳转到处理页面进行处理。后面会再配置一个控制权限的过滤器。

### 5.6.3　监听器

监听器（Listener）可以监听客户端的请求、服务端的操作等。通过监听器，可以自动激发一些操作，比如可以通过监听器实现在线的用户的计数。

当前，监听器可以对 request 对象、session 对象和 ServletContext（application）对象的创建、销毁、属性的添加、修改和删除事件进行监听。

与过滤器一样，监听器也是一个普通的 Java 类，只是该类需要实现一个或多个特定的接口，以实现对不同的对象进行监听。同样，监听器也需要配置到 web.xml 中。

监听器的配置与 Servlet、过滤器类似，但比它们要简单的多，监听器配置中没有 mapping 节点的配置，因为监听器处理的对象与 url-pattern 无关。Listener 配置只需要在 web.xml 中添加如下代码：

```xml
<listener>
    <listener-class>监听器类的包路径</listener-class>
</listener>
```

现在分别对不同类别监听器进行介绍。

1. 对 request 对象进行监听

对客户端请求的监听是 Servlet 2.4 规范中增加的技术。一旦我们可以监听客户端的请求，就可以对请求进行统一处理。比如做一个 Web 管理器程序，如果在本地或局域网内访问，就可以不用登录；如果是远程或外网访问那么就需要登录。要达到这个目的，我们就可以监听用户的请求，从用户请求中获取客户端的地址，并通过地址判断来做相应的处理。

要实现对 request 的监听，需要实现两个接口：javax.servlet.ServletRequestListener 和 javax.servlet.ServletRequestAttributeListener。前者主要对 request 的初始化和销毁事件进行监听，而后者主要对 request 的属性的添加、修改和删除事件进行监听。

ServletRequestListener 对象在对初始化事件和销毁事件进行监听时，都将接收一个 javax.servlet.ServletRequestEvent 类型的参数，我们可以通过该参数获得 Servlet 上下文和 Servlet 请求。

ServletRequestAttributeListener 在监听属性变化时，将获得一个 javax.servlet.ServletRequestAttributeEvent 类型的参数，该类扩展了 ServletRequestEvent 类，并添加了两个新方法：getName 和 getValue。getName 方法返回触发事件的属性的名称，getValue 返回属性的值。

一个 request 监听器实例如下。

（1）新建一个类 MyRequestListener，包名为：test.listeners，选择其实现的接口为上述的两个接口。

（2）在新创建的类的 requestInitialized 方法中添加以下代码：

```java
HttpServletRequest request = (HttpServletRequest)arg0.getServletRequest();
System.out.println("Request URL: " + request.getRequestURL());
System.out.println("Remote Port : " + request.getRemotePort());
System.out.println("Local Name : " + request.getLocalName());
System.out.println("Local Addr : " + request.getLocalAddr());
System.out.println("Local Port : " + request.getLocalPort());
```

（3）配置该监听器到 web.xml 中。在 web-app 上右击，选择 Add Child→context-param-login-config→listener，在 web.xml 中创建出相应的项，然后填写带完整包路径的类名即可。

在 web.xml 中添加如下配置代码：

```xml
<listener>
  <listener-class>test.listeners.MyRequestListener</listener-class>
</listener>
```

监听器运行的一个结果如图 5-15 所示。

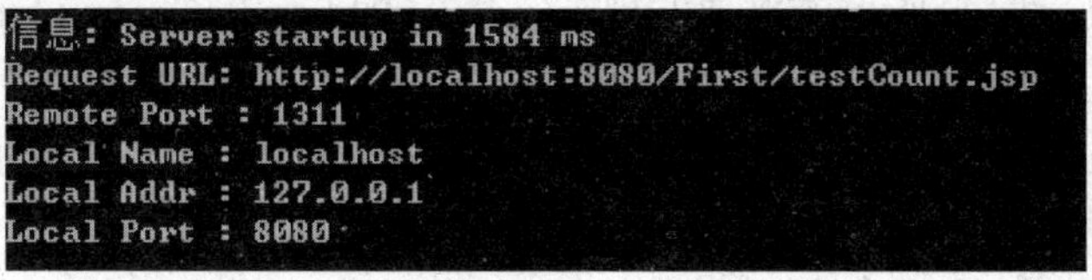
信息: Server startup in 1584 ms
Request URL: http://localhost:8080/First/testCount.jsp
Remote Port : 1311
Local Name : localhost
Local Addr : 127.0.0.1
Local Port : 8080

图 5-15　Request 监听器的运行结果

2. 对 session 对象进行监听

对 session 对象的监听主要是通过 3 个接口实现的，包括：HttpSessionAttributeListener、HttpSessionListener 和 HttpSessionActivationListener，下面分别介绍。

（1） HttpSessionAttributeListener 接口监听 HttpSession 中的属性的操作。当在 session 增加一个属性时，激发 attributeAdded(HttpSessionBindingEvent se) 方法；当在 session 删除一个属性时，激发 attributeRemoved(HttpSessionBindingEvent se)方法；当在 session 属性被重新设置时，激发 attributeReplaced(HttpSessionBindingEvent se) 方法。这和 ServletContextAttributeListener 比较类似。

HttpSessionBindingEvent 的主要方法如下。

java.lang.String getName()：回传属性的名称。

java.lang.Object getValue()：回传属性的值。

javax.servlet.http.HttpSession　getSession()：返回 session。

（2）HttpSessionListener 接口监听 HttpSession 的操作。当创建一个 session 时，激发 session Created(HttpSessionEvent se)方法；当销毁一个 session 时，激发 sessionDestroyed (HttpSessionEvent se)方法。

（3）HttpSessionActivationListener 接口主要用于同一个 session 转移至不同的 JVM 的情形。session 的 passivation 是指非活动的 session 被写入持久设备（比如硬盘），activate 是相反的过程。

3. 对 ServletContext 对象（application）进行监听

对 ServletContext 对象的监听主要是通过两个接口实现的，包括：ServletContextListener 和 ServletContextAttributeListener，下面分别介绍。

（1） ServletContextListener 用于监听 WEB 应用启动和销毁的事件，监听器类需要实现 javax.servlet.ServletContextListener 接口。

ServletContextListener 是 ServletContext 的监听者，如果 ServletContext 发生变化，如服务器启动时 ServletContext 被创建，服务器关闭时 ServletContext 将要被销毁。

ServletContextListener 接口的方法如下。

• void contextInitialized(ServletContextEvent sce)：通知正在接受的对象，应用程序已经被加载及初始化。

• void contextDestroyed(ServletContextEvent sce)：通知正在接受的对象，应用程序已经被载出。

ServletContextEvent 中的方法如下。

• ServletContext getServletContext()：取得 ServletContext 对象。

（2）ServletContextAttributeListener 用于监听 WEB 应用属性改变的事件，包括：增加属性、删除属性、修改属性，监听器类需要实现 javax.servlet.ServletContextAttributeListener 接口。
ServletContextAttributeListener 接口方法如下。

- void attributeAdded(ServletContextAttributeEvent scab)：若有对象加入 application 的范围，通知正在收听的对象。
- void attributeRemoved(ServletContextAttributeEvent scab)：若有对象从 application 的范围移除，通知正在收听的对象。
- void attributeReplaced(ServletContextAttributeEvent scab)：若在 Application 的范围中，有对象取代另一个对象时，通知正在收听的对象。

ServletContextAttributeEvent 中的方法如下。

- java.lang.String getName()：回传属性的名称。
- java.lang.Object getValue()：回传属性的值。

表 5-5 给出了常用的监听器接口及其事件类的列表。

表 5-5 常用的监听器接口及其事件类

| 监听器接口 | Event 类 |
| --- | --- |
| ServletContextListener | ServletContextEvent |
| ServletContextAttributeListener | ServletContextAttributeEvent |
| HttpSessionListener | HttpSessionEvent |
| HttpSessionActivationListener | |
| HttpSessionAttributeListener | HttpSessionBindingEvent |
| ServletRequestListener | ServletRequestEvent |
| ServletRequestAttributeListener | ServletRequestAttributeEvent |

**4. 监听器实例——BBS 在线用户计数器**

本实例创建一个监听器，实现对当前 BBS 所有访问用户和已登录用户的计数。我们已经知道，任何一个用户从开始访问 BBS 开始，服务器即为其分配一个 session；当 session 被销毁时，说明该用户已经退出了 BBS。因此我们可以通过对 session 对象的创建和销毁事件进行监听，以实现对访问 BBS 的用户进行计数。

当用户登录成功后，session 对象中将被写入一个属性（我们的程序写入的属性名为 logined），因此我们可以对此特定属性进行监听，如果该属性被创建，意味着一个用户已经登录系统；若该属性被销毁，说明用户已经退出该网站。

最后，要有一个合适的时机对访问网站的用户数和登录用户数进行初始化，这个操作我们可以在 Web 应用程序开始加载（刚启动）时实现，因此我们需要监听 ServletContext 的初始化操作。

为了在网页中能够使用记录的用户数量和登录用户数量，需要把它们添加为 application 对象的属性。

综上，我们将创建一个监听器，该监听器将实现 3 个接口：ServletContextListener、HttpSessionListener 和 HttpSessionAttributeListener。首先创建一个实现以上 3 个接口的类，其主要内容如清单 5-17 所示。

**清单 5-17 用户计数器代码**

```java
package test.listeners;

import javax.servlet.ServletContextEvent;
import javax.servlet.ServletContextListener;
import javax.servlet.http.HttpSessionAttributeListener;
import javax.servlet.http.HttpSessionBindingEvent;
import javax.servlet.http.HttpSessionEvent;
import javax.servlet.http.HttpSessionListener;

public class CounterListener implements HttpSessionListener,
        HttpSessionAttributeListener, ServletContextListener {

    int allUsers; // 对所有用户进行计数
    int loginedUser; // 对登录用户进行计数

    // 一个会话被创建，说明有个用户来访
    public void sessionCreated(HttpSessionEvent arg0) {
        allUsers++;
        arg0.getSession().getServletContext()
                .setAttribute("allUser",new Integer(allUsers));
    }

    // 会话被销毁，说明有个用户退出
    public void sessionDestroyed(HttpSessionEvent arg0) {
        allUsers--;
        arg0.getSession().getServletContext()
                .setAttribute("allUser",new Integer(allUsers));
    }

    // 有个属性被创建，判断是否是 logined 属性，若是，一个用户已经登录
    public void attributeAdded(HttpSessionBindingEvent arg0) {
        String name = arg0.getName();
        if ("logined".equals(name)){//一个登录用户已经登录成功
            loginedUser++;
            arg0.getSession().getServletContext()
                .setAttribute("logined",new Integer(loginedUser));
        }
    }

    // 有个属性被销毁，判断是否是 logined 属性，若是，一个用户已经退出
    public void attributeRemoved(HttpSessionBindingEvent arg0) {
        String name = arg0.getName();
        if ("logined".equals(name)){//一个登录用户已经登录成功
            loginedUser--;
```

```
            arg0.getSession().getServletContext()
                .setAttribute("logined",new Integer(loginedUser));
    }    }

    @Override
    public void attributeReplaced(HttpSessionBindingEvent arg0) {

    }

    @Override
    public void contextDestroyed(ServletContextEvent arg0) {

    }

    @Override
    public void contextInitialized(ServletContextEvent arg0) {
        loginedUser = 0;
        allUsers = 0;
    }

}
```

要使用监听器记录的用户数量数据，需要在网页中读取 application 对象的相应属性，代码如下：

```
out.println("All users: " + application.getAttribute("allUsers") + "<br>");
out.println("login users: " + application.getAttribute("loginUsers") + "<br>");
```

为测试计数监听器是不是能工作，我们继续使用用户登录的例子，在 loginProcess.jsp 页面中加入如下登录的判断：

```
if("admin".equals(userName) && "admin".equals(password)){
    session.setAttribute("logined", "true");
}
```

## 5.7 小  结

本章介绍了 Servlet 的基本概念，Servlet 与 JSP 之间的关系；讲述了 Servlet 的生命周期以及其提供的基本方法；举例说明了 Servlet 的开发过程；最后有又分别举例说明了 Servlet 与 JSP 的交互过程，应用 transfer 说明了 Servlet 获取表单数据的方法，应用 test 说明了从 Servlet 向 JSP 的信息传递的方法；介绍了 Servlet 的几个方面的高级应用，包括 Servlet 初始化参数的配置和使用方法，Servlet 过滤器的相关概念和使用以及各种监听器的使用方法。

## 5.8 习  题

1. 简述 Servlet 与 JSP 之间的联系与区别。
2. 简述 Servlet 的生命周期及方法。

3. 简述 Servlet 的开发步骤。
4. 简述 Servlet 与 JSP 之间的信息交互方式，并举例说明。
5. 请阐述 Servlet 过滤器的基本原理及其配置方法。
6. 如何创建一个简单的过滤器？
7. Servlet 的监听器可以实现什么功能？请给出一个实例。
8. 结合前面章节的学习，独立构建一个 Web 应用的登录过滤器，该应用的流程为：提供学生的注册页面，包括姓名、性别、年龄、学号、家庭住址等信息，在 Servlet 中获取这些信息，对需要保护的页面进行统一过滤。如果当前用户已经登录，则允许访问此页面，否则将用户请求跳转到 login.jsp，同时把该信息传递给另外一个 JSP 页面进行信息回显。

# 第 6 章 JavaBean 组件技术

本章主要介绍 JavaBeans 组件的概念和 JavaBeans 的编写和使用，了解 JavaBeans 编程约定，学会编写简单的 JavaBeans，掌握 JavaBeans 的配置，掌握应用 JSP 标记来使用 JavaBeans，掌握应用 JSP 和 JavaBeans 的开发模式。

## 6.1 JavaBean 概述

JavaBean 是为 Java 语言设计的软件组件模型，具有可重复使用和跨平台的特点。可以通过 JavaBean 来封装业务逻辑，进行数据库操作，从而很好地实现业务逻辑和前后台程序的分离。

JavaBean 其实就是一个简单的 Java 类，这也就意味着，Java 类的一切特征 JavaBean 也都具有。JavaBean 同样可以使用封装、继承、多态等特性。

JavaBean 可以分为两类，一类是有用户接口（UI）的 JavaBean，一类是没有用户接口的 JavaBean。一般在 JSP 中使用的都是没有用户接口的 JavaBean，因此本章所介绍的 JavaBean 都是指没有用户接口的 JavaBean。这类 JavaBean 只是简单地进行业务封装，如数据运算和处理、数据库操作等。

一个标准的 JavaBean 应该具有如下几个特点：

（1）JavaBean 必须是一个公开的类，也就是说 JavaBean 的类访问权限必须是 public。

（2）JavaBean 必须具有一个无参数的构造方法。如果在 JavaBean 中定义了自定义的有参构造方法，就必须添加一个无参数构造方法，否则将无法设置属性；如果没有定义自定义的有参构造方法，则可以利用编译器自动添加的无参构造方法。

（3）JavaBean 一般将属性设置成私有的，通过使用 getXXX()方法和 setXXX()方法来进行属性的取得和设置。

## 6.2 JavaBean 的编写和使用

实际上，JavaBean 是对遵循指定的编码约定的 Java 类的一种别称。从技术上讲，任何 Java 类，如果实现了 java.io.Serializable 接口并且提供默认构造方法（没有参数的构造方法），就可以成为 JavaBean。

### 6.2.1 编写 JavaBean

JavaBean 模型如下：

```
package 包名;
public class 类名 {
    构造方法 ();
    属性 (Property);
```

```
        方法 (Method);
    }
```

编写 JavaBean 就是编写一个 Java 的类，这个类创建的一个对象称作一个 beans。
Javabean 类中的方法如下：
类的成员名字是 XXX，获取和更改 XXX 形式如下：
getXXX()：获取属性 XXX。
setXXX()：修改属性 XXX。
boolean 型属性，允许用 is 代替 get 和 set。
类中方法的访问属性必须是 public。
类中构造方法必须是 public，并且是无参数。
JavaBean 属性值的设定方法如下：

```
public void set 设定方法名称（数据类型 参数）{
    this.变量=参数;
}
```

例 6-1：User.java

```java
public class User
{
    private int userid;
    private int userclass;
    private String username;
    private String password;
    private String useremail;

    public int getUserid() {
        return userid;
    }

    public void setUserid(int userid) {
        this.userid = userid;
    }

    public int getUserclass() {
        return userclass;
    }

    public void setUserclass(int userclass) {
        this.userclass = userclass;
    }

    public String getUsername() {
        return username;
    }

    public void setUsername(String username) {
```

```java
        this.username = username;
    }

    public String getPassword() {
        return password;
    }

    public void setPassword(String password) {
        this.password = password;
    }

    public String getUseremail() {
        return useremail;
    }

    public void setUseremail(String useremail) {
        this.useremail = useremail;
    }
}
```

示例代码声明该类的访问权限为 public，符合 JavaBean 定义的第一个特点。代码中没有定义自定义的有参构造方法，这时候可以利用编译器自动添加的无参构造方法，符合 JavaBean 定义的第二个特点。代码属性访问级别为 private，然后通过添加相应的 getXXX()方法和 setXXX()方法来进行属性的取得和设置，符合 JavaBean 定义的第三个特点。

### 6.2.2 使用 JavaBean

在 JSP 中调用 JavaBean 有如下两个优点：

（1）提高代码的可复用性。对于通常使用的业务逻辑代码，如数据运算和处理、数据库操作等，可以封装到 JavaBean 中。在 JSP 文件中可以多次调用 JavaBean 中的方法来实现快速的程序开发。

（2）将 HTML 代码和 Java 代码进行分离，利于程序开发维护。将业务逻辑进行封装，使得业务逻辑代码和显示代码相分离，不会互相干扰，避免了代码又多又复杂的问题，方便了日后的维护。

要想在 JSP 中调用 JavaBean，就需要使用<jsp:useBean>动作指令，该动作指令主要用于创建和查找 JavaBean 的示例对象。其语法格式如下：

```
<jsp:useBean id="对象名称" scope="存储范围" class="类名"></jsp:useBean>
```

其中 ID 属性表示该 JavaBean 实例化后的对象名称；scope 属性用来指定该 JavaBean 的范围，也就是指 JavaBean 实例化后的对象存储范围；范围的取值分别是 page、request、session 和 application；class 属性用来指定 JavaBean 的类名，这里所指的类名包括包名和类名。

例 6-2：在 JSP 中使用 JavaBean。

```
<%@ page language="java" contentType="text/html;charset=utf-8"%>
<%--通过 useBean 动作指令调用 JavaBean--%>
<jsp:useBean id="user" scope="page" class="net.bean.User"></jsp:useBean>
```

```
<html>
 <head>
  <title>调用 JavaBean</title>
 </head>
 <body>
 <%
   // 设置 user 的 username 属性
   user.setUsername("zhangsan");
   // 设置 user 的 password 属性
   user.setPassword("1234");
   // 打印输出 user 的 username 属性
   out.println("用户名: " + user.getUsername() + "<br>");
     // 打印输出 user 的 password 属性
   out.println("用户密码: " + user.getPassword());
 %>
 </body>
</html>
```

在 JSP 页面通过<jsp:useBean>指令调用名为 User 的 JavaBean，并设置其实例化对象名为 user，声明其存储范围为 page 范围。在 Java 片段中通过 user 对象设置其属性值，并通过 user 实例化对象获得其属性值，然后输出在页面上。

### 6.2.3 设置 JavaBean 属性

JSP 中提供了一个<jsp:setProperty>动作指令来设置 JavaBean 属性，其有如下 4 种语法格式：

```
<jsp:setProperty name="实例化对象名" property="*"/>
<jsp:setProperty name="实例化对象名" property="属性名称"/>
<jsp:setProperty name="实例化对象名" property="属性名称" param="参数名称"/>
<jsp:setProperty name="实例化对象名" property="属性名称" value="属性值" />
```

其中 name 属性使用设置实例化对象名，和<jsp:useBean>中的 id 属性保持一致，property 属性用来指定 JavaBean 属性名称，param 属性用来指定接收参数名称，value 属性用来指定属性值。

下面分别介绍这 4 种<jsp:setProperty>动作指令的作用。

1. 根据所有参数设置 JavaBean 属性

<jsp:setProperty>动作指令用来根据所有参数设置 JavaBean 属性，其语法格式如下：

```
<jsp:setProperty name="实例化对象名" property="*"/>
```

其中"*"表示根据表单传递的所有参数来设置 JavaBean 的属性，JSP 引擎就会把所有 request 参数与 JavaBean 中的 setXXX 方法名进行匹配，即所有的名称与 bean 的属性匹配的 request 参数都将被传递到相应对象的属性中。比如通过表单传递了两个参数，如 username 和 password，这时就可以自动地对 JavaBean 中的 username 属性及 password 属性进行赋值。这里必须注意的是，表单的参数值必须和 JavaBean 中的属性名称保持大小写一致，否则无法进行赋值操作。

如果不用< jsp:setProperty>，则需要做大量的类型转换工作，利用 jsp:setProperty 可以实现自动类型转换。

**例 6-3**：根据所有提交表单参数设置 JavaBean 属性的示例。该例包含两个文件，一个是用来传递表单参数的 JavaBean 数据表单 parameterForm.jsp，一个是设置 JavaBean 属性页面 setProperty1.jsp。

parameterForm.jsp

```jsp
<%@ page language="java" import="java.util.*" pageEncoding="utf-8"%>
<%
String path = request.getContextPath();
String basePath = request.getScheme()+"://"+request.getServerName()+":"
    +request.getServerPort()+path+"/";
%>

<!DOCTYPE HTML PUBLIC "-//W3C//DTD HTML 4.01 Transitional//EN">
<html>
  <head>
    <base href="<%=basePath%>">

    <title>My JSP 'userForm.jsp' starting page</title>

    <meta http-equiv="pragma" content="no-cache">
    <meta http-equiv="cache-control" content="no-cache">
    <meta http-equiv="expires" content="0">
    <meta http-equiv="keywords" content="keyword1,keyword2,keyword3">
    <meta http-equiv="description" content="This is my page">
    <!--
    <link rel="stylesheet" type="text/css" href="styles.css">
    -->

  </head>

  <body>
    <form action="setProperty3.jsp" method="post">
    <table>
        <tr> <td colspan="2">JavaBean 数据表单</td></tr>
        <tr>
             <td>用户名：</td>
             <td><input type="text" name="username"></td>
        </tr>
        <tr>
             <td>密  码：</td>
             <td><input type="password" name="password"></td>
        </tr>
        <tr>
             <td colspan="2">
                 <input type="submit" value="提交">
                 <input type="reset" value="重置">
             </td>
        </tr>
```

```
        </table>
    </form>

  </body>
</html>
```

**setProperty.jsp**

```jsp
<%@ page language="java" import="java.util.*" pageEncoding="utf-8"%>
<jsp:useBean id="user" scope="page" class="net.bean.User"/>
<%
String path = request.getContextPath();
String basePath = request.getScheme()+"://"+request.getServer-
    Name()+":"+request.getServerPort()+path+"/";
%>

<!DOCTYPE HTML PUBLIC "-//W3C//DTD HTML 4.01 Transitional//EN">
<html>
  <head>
    <base href="<%=basePath%>">

    <title>My JSP 'setProperty.jsp' starting page</title>

    <meta http-equiv="pragma" content="no-cache">
    <meta http-equiv="cache-control" content="no-cache">
    <meta http-equiv="expires" content="0">
    <meta http-equiv="keywords" content="keyword1,
                    keyword2,keyword3">
    <meta http-equiv="description" content="This is my page">
    <!--
    <link rel="stylesheet" type="text/css" href="styles.css">
    -->

  </head>

  <body>
    <%--根据传递的所有参数设置 JavaBean 中属性 --%>
      <jsp:setProperty name="user" property="*"/>

      <!-- 打印输出 user 的 username 属性-->
      用户名：<%=user.getUsername()%><br>
      <!-- 打印输出 user 的 password 属性  -->
      密  码：<%=user.getPassword()%>
  </body>
</html>
```

填写用户表单信息，如用户名为 root，密码为 123，填写完成后单击"提交"按钮进行提交。页面跳转到 setProperty.jsp 页面，如图 6-1 所示。

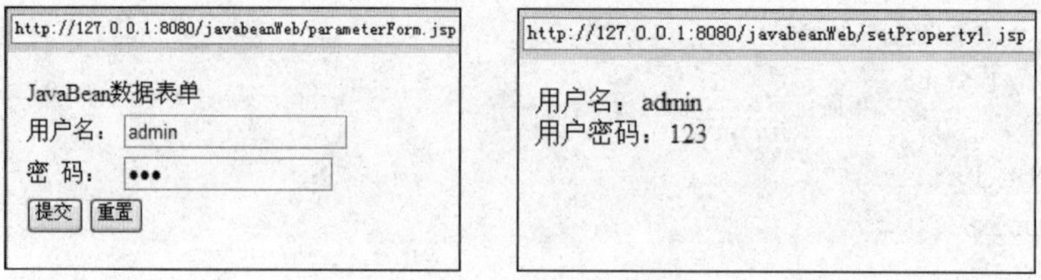

图 6-1 根据传递的所有参数设置 JavaBean 中属性

例子中通过<jsp:useBean>指令调用名为 UserBean 的 JavaBean，并设置其实例化对象名为 user，其存储范围为 page 范围。通过<jsp:setProperty>动作指令来根据所有参数设置 JavaBean 属性，通过实例化对象名 user 分别获得其属性值并输出到页面上。

2. 根据指定参数设置 JavaBean 属性

<jsp:setProperty>动作指令用来根据指定参数设置 JavaBean 属性，其语法格式如下：

```
<jsp:setProperty name="实例化对象名" property="数值名称"/>
```

与上一个动作指令相比，这个<jsp:setProperty>动作指令具有更好的弹性。上一个<jsp:setProperty>动作指令要求设置所有的参数，而这个<jsp:setProperty>动作指令可以用来设置指定的参数。比如通过表单传递了两个参数，如 username 和 password，这时就可以指定只为 JavaBean 的 username 属性赋值，也可以指定只为 JavaBean 的 password 属性赋值。

例 6-4：代码段 setProperty2.jsp 根据指定参数设置 JavaBean 属性。

```jsp
<%@ page language="java" import="java.util.*" pageEncoding="utf-8"%>
<%
String path = request.getContextPath();
String basePath = request.getScheme()+"://"+request.getServer-
    Name()+":"+request.getServerPort()+path+"/";
%>

<!DOCTYPE HTML PUBLIC "-//W3C//DTD HTML 4.01 Transitional//EN">
<html>
  <head>
    <base href="<%=basePath%>">
    
    <title>My JSP 'setProperty2.jsp' starting page</title>
    
    <meta http-equiv="pragma" content="no-cache">
    <meta http-equiv="cache-control" content="no-cache">
    <meta http-equiv="expires" content="0">
    <meta http-equiv="keywords" content="keyword1,keyword2,keyword3">
    <meta http-equiv="description" content="This is my page">
    <!--
    <link rel="stylesheet" type="text/css" href="styles.css">
    -->

  </head>
```

```
        <body>
            <%--通过 useBean 动作指定调用 JavaBean --%>
        <jsp:useBean id="user" scope="page"
class="net.bean.User"></jsp:useBean>
        <%--设置 username 属性值 --%>
        <jsp:setProperty name="user" property="username"/>

        <!-- 打印输出 user 的 username 属性-->
        用户名：<%=user.getUsername()%><br>
        <!-- 打印输出 user 的 password 属性  -->
        密  码：<%=user.getPassword()%>

        </body>
</html>
```

修改其中 JavaBean 数据表单的提交页面，设置页面提交到 setProperty2.jsp。

填写用户表单信息，如用户名为 admin，密码为 123，填写完成后单击"提交"按钮进行提交。页面跳转到 setProperty2.jsp 页面，如图 6-2 所示。

图 6-2　根据指定参数设置 JavaBean 中属性

例子代码通过<jsp:useBean>指令调用名为 UserBean 的 JavaBean，并设置其实例化对象名为 user，其存储范围为 page 范围。代码通过<jsp:setProperty>动作指令来根据指定参数 username 设置 JavaBean 属性。代码通过实例化对象名 user 分别获得其属性值并输出到页面上。因为并没有指定参数 password 来设置 JavaBean 中的属性，所以其值为 null。

3. 根据指定参数设置指定 JavaBean 属性

用来根据指定参数设置指定 JavaBean 属性，其语法格式如下：

```
<jsp:setProperty name="实例化对象名" property="属性名称" param="参数名称"/>
```

与前两种<jsp:setProperty>动作指令相比，该种动作指令更加具有弹性。需要设置参数和 JavaBean 属性必须相同，而且必须保证大小写一致。而<jsp:setProperty>动作指令没有此限制，因为可以通过其指定需要设置的 JavaBean 属性。

例 6-5：根据指定参数设置指定 JavaBean 属性。包含两个文件，一个是用来传递参数的 JavaBean 数据表单 parameterForm.jsp，一个是设置 JavaBean 属性页面 setProperty3.jsp。

parameterForm.jsp

```
<form action="setProperty3.jsp" method="post">
    <table>
        <tr> <td colspan="2">JavaBean 数据表单</td></tr>
        <tr>
```

```html
                <td>用户名：</td>
                <td><input type="text" name="username"></td>
        </tr>
        <tr>
                <td>密  码：</td>
                <td><input type="password" name="password"></td>
        </tr>
        <tr>
                <td colspan="2">
                    <input type="submit" value="提交">
                    <input type="reset" value="重置">
                </td>
        </tr>
    </table>
</form>
```

### setProperty3.jsp

```jsp
<%@ page language="java" import="java.util.*" pageEncoding="utf-8"%>
<jsp:useBean id="user" scope="page" class="net.bean.User"/>
<%
String path = request.getContextPath();
String basePath = request.getScheme()+"://"+request.getServerName()+":"
   +request.getServerPort()+path+"/";
%>

<!DOCTYPE HTML PUBLIC "-//W3C//DTD HTML 4.01 Transitional//EN">
<html>
  <head>
    <base href="<%=basePath%>">

    <title>My JSP 'setProperty3.jsp' starting page</title>

    <meta http-equiv="pragma" content="no-cache">
    <meta http-equiv="cache-control" content="no-cache">
    <meta http-equiv="expires" content="0">
    <meta http-equiv="keywords" content="keyword1,keyword2,keyword3">
    <meta http-equiv="description" content="This is my page">
    <!--
    <link rel="stylesheet" type="text/css" href="styles.css">
    -->

  </head>

  <body>
    <%--设置 username 属性，其值为 username 参数值--%>
    <jsp:setProperty name="user" property="username" param="username"/>
    <%--设置 password 属性，其值为 password 参数值--%>
    <jsp:setProperty name="user" property="password" param="password"/>
```

```
        用户名：<%=user.getUsername()%><br>
        密  码：<%=user.getPassword()%>

    </body>
</html>
```

修改 JavaBean 数据表单的提交页面，设置页面提交到 setProperty3.jsp，并修改密码输入框的 name 属性，设置其为 userpassword。

填写用户表单信息，如用户名为 admin，密码为 123，填写完成后单击"提交"按钮进行提交。页面跳转到 setProperty3.jsp 页面，如图 6-3 所示。

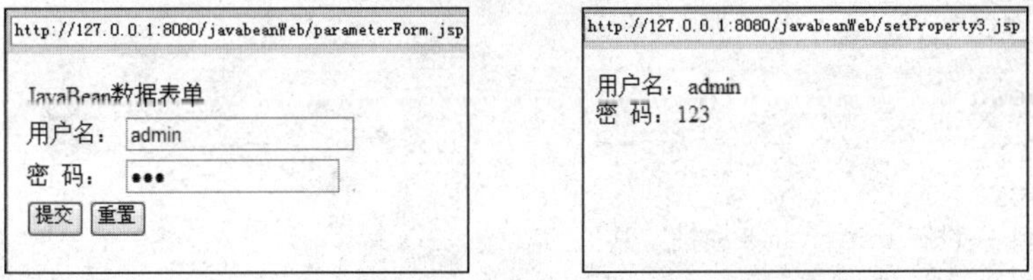

图 6-3　根据指定参数设置指定 JavaBean 属性

示例代码通过<jsp:useBean>指令调用名为 UserBean 的 JavaBean，并设置其实例化对象名为 user，其存储范围为 page 范围。代码通过<jsp:setProperty>动作指令来根据参数 username 设置 JavaBean 的 username 属性，通过<jsp:setProperty>动作指令来根据参数 userpassword 设置 JavaBean 的 password 属性，通过实例化对象名 user 分别获得其属性值并输出在页面上。

4. 设置指定 JavaBean 属性为指定值

用来设置指定 JavaBean 属性为指定值，其语法格式如下：

```
<jsp:setProperty name="实例化对象名" property="属性名称" value="属性值" />
```

这个<jsp:setProperty>动作指令相比前 3 个<jsp:setProperty>动作指令更加具有弹性。前面 3 种<jsp:setProperty>动作指令都需要接收表单参数，而这第 4 种可以根据需要动态地设置 JavaBean 属性值。

**例 6-6**：根据指定参数设置指定 JavaBean 属性的示例。

setProperty4.jsp

```
<%@ page language="java" import="java.util.*" pageEncoding="utf-8"%>
<%--通过 useBean 动作指定调用 JavaBean --%>
<jsp:useBean id="user" scope="page" class="net.bean.User"/>
<%
String path = request.getContextPath();
String basePath = request.getScheme()+"://"+request.getServerName()+":"
    +request.getServerPort()+path+"/";
%>

<!DOCTYPE HTML PUBLIC "-//W3C//DTD HTML 4.01 Transitional//EN">
<html>
    <head>
        <base href="<%=basePath%>">
```

```
        <title>My JSP 'setProperty4.jsp' starting page</title>

        <meta http-equiv="pragma" content="no-cache">
        <meta http-equiv="cache-control" content="no-cache">
        <meta http-equiv="expires" content="0">
        <meta http-equiv="keywords" content="keyword1,keyword2,keyword3">
        <meta http-equiv="description" content="This is my page">
        <!--
        <link rel="stylesheet" type="text/css" href="styles.css">
        -->

    </head>

    <body>
        <%--设置 username 属性,其值为 abc --%>
        <jsp:setProperty name="user" property="username" value="admin" />
        <%--设置 password 属性,其值为 123--%>
        <jsp:setProperty name="user" property="password" value="123"/>

            <!-- 打印输出 user 的 username 属性-->
        用户名:<%=user.getUsername()%><br>
            <!-- 打印输出 user 的 password 属性  -->
        密  码: <%=user.getPassword()%>
    </body>
</html>
```

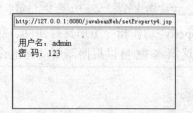

图 6-4 设置指定 JavaBean 属性为指定值

示例代码通过<jsp:useBean>指令调用名为 UserBean 的 JavaBean,并设置其实例化对象名为 user,其存储范围为 page 范围。代码通过<jsp:setProperty>动作指令来设置 JavaBean 的 username 属性值为 admin,通过<jsp:setProperty>动作指令来设置 JavaBean 的 password 属性为 123,通过实例化对象名 user 分别获得其属性值并输出在页面上,如图 6-4 所示。

### 6.2.4 获得 JavaBean 属性

前面介绍的 JavaBean 属性都是通过调用实例化对象名从而获得的,下面来介绍一种更加简便的方法。

JSP 提供了一个<jsp:getProperty>动作指令,用来很方便地获得 JavaBean 属性,其语法格式如下:

<jsp:getProperty name="实例化对象名" property="属性名称"/>

其中 name 属性用来设置实例化对象名,同样必须和<jsp:useBean>中的 id 属性保持一致;property 属性用来指定需要获得的 JavaBean 属性名称。

**例 6-7**：使用<jsp:getProperty>动作指令获得 JavaBean 属性的示例。

getProperty.jsp

```jsp
<%@ page language="java" import="java.util.*" pageEncoding="utf-8"%>
<%
String path = request.getContextPath();
String basePath = request.getScheme()+"://"+request.getServerName()+":"
    +request.getServerPort()+path+"/";
%>

<!DOCTYPE HTML PUBLIC "-//W3C//DTD HTML 4.01 Transitional//EN">
<html>
  <head>
    <base href="<%=basePath%>">

    <title>My JSP 'getProperty.jsp' starting page</title>

    <meta http-equiv="pragma" content="no-cache">
    <meta http-equiv="cache-control" content="no-cache">
    <meta http-equiv="expires" content="0">
    <meta http-equiv="keywords" content="keyword1,keyword2,keyword3">
    <meta http-equiv="description" content="This is my page">
    <!--
    <link rel="stylesheet" type="text/css" href="styles.css">
    -->

  </head>

  <body>
    <%--通过 userBean 动作指定使用 UserBean --%>
    <jsp:useBean id="user" scope="page" class="net.bean.User"></jsp:useBean>
    <%--设置 username 属性，其值为 admin --%>
    <jsp:setProperty name="user" property="username" value="admin"/>
    <%--设置 password 属性，其值为 123 --%>
    <jsp:setProperty name="user" property="password" value="123"/>
        <%--获得 username 属性 --%>
        <jsp:getProperty name="user" property="username"/>
    <%--获得 password 属性--%>
    <jsp:getProperty name="user" property="password"/>

  </body>
</html>
```

代码通过<jsp:useBean>指令调用名为 UserBean 的 JavaBean，并设置其实例化对象名为 user，其存储范围为 page 范围。代码通过<jsp:setProperty>动作指令来设置 JavaBean 的 username 属性值为 admin，通过<jsp:setProperty>动作指令来设置 JavaBean 的 password 属性为 admin。代码通过<jsp:getProperty>动作指令获得 JavaBean 的 username 属性和 password 属性，如图 6-5 所示。

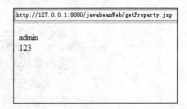

图 6-5 获得 JavaBean 属性

## 6.2.5 设置 JavaBean 的范围

JSP 属性有 4 种存储范围，分别为 page、request、session 及 application。同样也可以设置 JavaBean 的存储范围，其取值和意义同属性保存范围完全相同。下面分别介绍如何设置不同范围的 JavaBean，以及不同存储范围的区别。

**1. 设置 page 范围的 JavaBean**

page 范围的 JavaBean，设置的 JavaBean 只在当前页面有效。

例 6-8：设置 page 范围的 JavaBean 示例。该范例包含两个文件，一个是用来设置 JavaBean 属性的 pageBean.jsp 页面，一个是跳转后获得 JavaBean 属性的 getPageBean.jsp 页面。

pageBean.jsp

```jsp
<%@ page language="java" import="java.util.*" pageEncoding="utf-8"%>
<%--通过 useBean 动作指定调用 JavaBean --%>
<jsp:useBean id="user" scope="page" class="net.bean.User"></jsp:useBean>
<%
String path = request.getContextPath();
String basePath = request.getScheme()+"://"+request.getServerName()+":"
    +request.getServerPort()+path+"/";
%>

<!DOCTYPE HTML PUBLIC "-//W3C//DTD HTML 4.01 Transitional//EN">
<html>
  <head>
    <base href="<%=basePath%>">

    <title>My JSP 'pageBean.jsp' starting page</title>

    <meta http-equiv="pragma" content="no-cache">
    <meta http-equiv="cache-control" content="no-cache">
    <meta http-equiv="expires" content="0">
    <meta http-equiv="keywords" content="keyword1,keyword2,keyword3">
    <meta http-equiv="description" content="This is my page">
    <!--
    <link rel="stylesheet" type="text/css" href="styles.css">
    -->

  </head>

  <body>
    <%--设置 username 属性，其值为 admin --%>
    <jsp:setProperty name="user" property="username" value="admin"/>
    <%--设置 password 属性，其值为 admin --%>
    <jsp:setProperty name="user" property="password" value="admin"/>
    <%--服务器端跳转到 getPageBean.jsp --%>
    <jsp:forward page="getPageBean.jsp"></jsp:forward>

  </body>
</html>
```

getPageBean.jsp

```jsp
<%@ page language="java" import="java.util.*" pageEncoding="utf-8"%>
<%--通过 useBean 动作指定调用 JavaBean --%>
<jsp:useBean id="user" scope="page" class="net.bean.User"></jsp:useBean>
<%
String path = request.getContextPath();
String basePath = request.getScheme()+"://"+request.getServerName()+":"
   +request.getServerPort()+path+"/";
%>

<!DOCTYPE HTML PUBLIC "-//W3C//DTD HTML 4.01 Transitional//EN">
<html>
  <head>
    <base href="<%=basePath%>">
    
    <title>My JSP 'getPageBean.jsp' starting page</title>
    
    <meta http-equiv="pragma" content="no-cache">
    <meta http-equiv="cache-control" content="no-cache">
    <meta http-equiv="expires" content="0">
    <meta http-equiv="keywords" content="keyword1,keyword2,keyword3">
    <meta http-equiv="description" content="This is my page">
    <!--
    <link rel="stylesheet" type="text/css" href="styles.css">
    -->

  </head>
  
  <body>
    <%--获得 username 属性--%>
    用户名：<jsp:getProperty name="user" property="username"/><br>
    <%--获得 password 属性--%>
    密  码：<jsp:getProperty name="user" property="password"/>
    
  </body>
</html>
```

pageBean 页面通过<jsp:useBean>指令调用名为 UserBean 的 JavaBean，并设置其实例化对象名为 user，其存储范围为 page 范围。通过<jsp:setProperty>设置 JavaBean 的 username 属性值为 admin，设置 JavaBean 的 password 属性值为 admin。然后通过<jsp:forward>设置页面跳转到 getPageBean.jsp。getPageBean 页面通过<jsp:getProperty>获得 JavaBean 的 username 属性和 password 属性。因为 JavaBean 的存储范围为 page，所以在其他页面无法获得，所以在页面上输出其值 null，如图 6-6 所示。

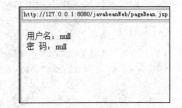

图 6-6  设置 page 范围的 JavaBean

## 2. 设置 request 范围的 JavaBean

request 范围的 JavaBean，设置的 JavaBean 在一次请求范围内有效。如果页面从一个页面跳转到另一个页面，那么该属性也就失效了。这里所指的跳转是指客户端跳转，比如客户单击超链接跳转到其他页面，或者通过浏览器地址栏浏览其他网页。如果使用服务器端跳转 <jsp:forward>，该 JavaBean 仍然有效。

**例 6-9**：设置 request 范围的 JavaBean 示例。该范例包含两个文件，一个是用来设置 JavaBean 属性的 requestBean.jsp 页面，一个是跳转后获得 JavaBean 属性的 getRequestBean.jsp 页面。

requestBean.jsp

```jsp
<%@ page language="java" import="java.util.*" pageEncoding="utf-8"%>
<%--通过useBean动作指定调用JavaBean --%>
<jsp:useBean id="user" scope="page" class="net.bean.User"></jsp:useBean>
<%
String path = request.getContextPath();
String basePath = request.getScheme()+"://"+request.getServerName()+":"
    +request.getServerPort()+path+"/";
%>

<!DOCTYPE HTML PUBLIC "-//W3C//DTD HTML 4.01 Transitional//EN">
<html>
  <head>
    <base href="<%=basePath%>">

    <title>My JSP 'pageBean.jsp' starting page</title>

    <meta http-equiv="pragma" content="no-cache">
    <meta http-equiv="cache-control" content="no-cache">
    <meta http-equiv="expires" content="0">
    <meta http-equiv="keywords" content="keyword1,keyword2,keyword3">
    <meta http-equiv="description" content="This is my page">
    <!--
    <link rel="stylesheet" type="text/css" href="styles.css">
    -->

  </head>

  <body>
    <%--设置username属性，其值为request_username --%>
    <jsp:setProperty name="user" property="username"
        value="request_ username" />
    <%--设置password属性，其值为request_password --%>
    <jsp:setProperty name="user" property="password"
        value="request_ password" />

    <%--服务器端跳转到getRequestBean.jsp.jsp --%>
    <jsp:forward page="getRequestBean.jsp"></jsp:forward>
  </body>
</html>
```

getRequestBean.jsp

```jsp
<%@ page language="java" import="java.util.*" pageEncoding="utf-8"%>
<%--通过useBean动作指定调用JavaBean --%>
<jsp:useBean id="user" scope="page" class="net.bean.User"></jsp:useBean>
<%
String path = request.getContextPath();
String basePath = request.getScheme()+"://"+request.getServerName()+":"
   +request.getServerPort()+path+"/";
%>

<!DOCTYPE HTML PUBLIC "-//W3C//DTD HTML 4.01 Transitional//EN">
<html>
  <head>
    <base href="<%=basePath%>">
    
    <title>My JSP 'getPageBean.jsp' starting page</title>
    
	<meta http-equiv="pragma" content="no-cache">
	<meta http-equiv="cache-control" content="no-cache">
	<meta http-equiv="expires" content="0">
	<meta http-equiv="keywords" content="keyword1,keyword2,keyword3">
	<meta http-equiv="description" content="This is my page">
	<!--
	<link rel="stylesheet" type="text/css" href="styles.css">
	-->

  </head>
  
  <body>
    <%--获得username属性--%>
    用户名：<jsp:getProperty name="user" property="username"/><br>
    <%--获得password属性--%>
    密  码：<jsp:getProperty name="user" property="password"/>
    
  </body>
</html>
```

requestBean.jsp 通过<jsp:useBean>指令调用名为 UserBean 的 JavaBean，并设置其实例化对象名为 user，其存储范围为 request 范围。通过<jsp:setProperty>设置 JavaBean 的 username 属性值为 request_username，设置 JavaBean 的 password 属性值为 request_password。通过<jsp:forward>设置页面跳转到 RequestJavaBeanDemo2.jsp。RequestJavaBeanDemo2.jsp 通过 <jsp:getProperty> 获得 JavaBean 的 username 属性和 password 属性。因为 JavaBean 的存储范围为 request，所以如果是服务器端跳转的话，可以获得其属性值，如图 6-7 所示。

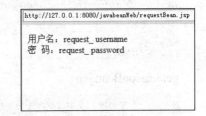

图 6-7 设置 request 范围的 JavaBean

## 3. 设置 session 范围的 JavaBean

session 范围 JavaBean，是客户浏览器与服务器一次会话期间范围内，如果和服务器断开连接，那么这个 JavaBean 也就失效了。

**例6-10**：设置 session 范围的 JavaBean 示例。该范例包含两个文件，一个是用来设置 JavaBean 属性的 sessionBean.jsp 页面，一个是跳转后获得 JavaBean 属性的 getSessionBean.jsp 页面。

sessionBean.jsp

```jsp
<%@ page language="java" import="java.util.*" pageEncoding="utf-8"%>
<%--通过 useBean 动作指定调用 JavaBean --%>
<jsp:useBean id="user" scope="request" class="net.bean.User"></jsp:useBean>
<%
String path = request.getContextPath();
String basePath = request.getScheme()+"://"+request.getServerName()+":"
   +request.getServerPort()+path+"/";
%>

<!DOCTYPE HTML PUBLIC "-//W3C//DTD HTML 4.01 Transitional//EN">
<html>
  <head>
    <base href="<%=basePath%>">

    <title>My JSP 'sessionBean.jsp' starting page</title>

    <meta http-equiv="pragma" content="no-cache">
    <meta http-equiv="cache-control" content="no-cache">
    <meta http-equiv="expires" content="0">
    <meta http-equiv="keywords" content="keyword1,keyword2,keyword3">
    <meta http-equiv="description" content="This is my page">
    <!--
    <link rel="stylesheet" type="text/css" href="styles.css">
    -->

  </head>

  <body>
    <%--设置 username 属性，其值为 session_username --%>
    <jsp:setProperty name="user" property="username" value="session_username"/>
    <%--设置 password 属性，其值为 session_password --%>
    <jsp:setProperty name="user" property="password" value="session_password"/>
    <%--跳转到 getSessionBean.jsp --%>
    <a href="getSessionBean.jsp">跳转到 getSessionBean.jsp</a>

  </body>
</html>
```

getSessionBean.jsp

```jsp
<%@ page language="java" import="java.util.*" pageEncoding="utf-8"%>
<%--通过 useBean 动作指定调用 JavaBean --%>
<jsp:useBean id="user" scope="request" class="net.bean.User"></jsp:useBean>
<%
```

```jsp
	String path = request.getContextPath();
	String basePath = request.getScheme()+"://"+request.getServerName()+":"
		+request.getServerPort()+path+"/";
%>

<!DOCTYPE HTML PUBLIC "-//W3C//DTD HTML 4.01 Transitional//EN">
<html>
  <head>
    <base href="<%=basePath%>">
    
    <title>My JSP 'getSessionBean.jsp' starting page</title>
    
	<meta http-equiv="pragma" content="no-cache">
	<meta http-equiv="cache-control" content="no-cache">
	<meta http-equiv="expires" content="0">    
	<meta http-equiv="keywords" content="keyword1,keyword2,keyword3">
	<meta http-equiv="description" content="This is my page">
	<!--
	<link rel="stylesheet" type="text/css" href="styles.css">
	-->

  </head>
  
  <body>
       <%--获得username属性--%>
       用户名:<jsp:getProperty name="user" property="username"/><br>
       <%--获得password属性--%>
       密  码:<jsp:getProperty name="user" property="password"/>
       
  </body>
</html>
```

sessionBean.jsp 通过<jsp:useBean>指令调用名为 UserBean 的 JavaBean，并设置其实例化对象名为 user，其存储范围为 session 范围。通过<jsp:setProperty>设置 JavaBean 的 username 属性值为 session_username，设置 JavaBean 的 password 属性值为 session_password。通过<jsp:forward>设置页面跳转到 getSessionBean.jsp 通过< jsp:getProperty>获得 JavaBean 的 username 属性和 password 属性。因为 JavaBean 的存储范围为 session，因此无论是客户端跳转还是服务器端跳转，都能够获得 session 范围的 JavaBean。但是如果重新打开浏览器，就不能获得 session 范围的 JavaBean，因为会话已经结束，如图 6-8 所示。

 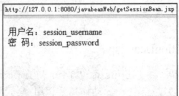

图 6-8　设置 session 范围的 JavaBean

## 4. 设置 application 范围的 JavaBean

application 范围，指在整个服务器运行期间范围，直到服务器停止以后才会失效。

**例 6-11**：设置 application 范围的 JavaBean 示例。该范例包含两个文件，一个是用来设置 JavaBean 属性的 applicationBean.jsp 页面，一个是跳转后获得 JavaBean 属性的 getApplicationBean.jsp 页面。

applicationBean.jsp

```jsp
<%@ page language="java" import="java.util.*" pageEncoding="utf-8"%>
<%--通过 useBean 动作指定调用 JavaBean --%>
<jsp:useBean id="user" scope="application" class="net.bean.User"></jsp:useBean>
<%
String path = request.getContextPath();
String basePath = request.getScheme()+"://"+request.getServerName()+":"
    +request.getServerPort()+path+"/";
%>

<!DOCTYPE HTML PUBLIC "-//W3C//DTD HTML 4.01 Transitional//EN">
<html>
  <head>
    <base href="<%=basePath%>">

    <title>My JSP 'applicationBean.jsp' starting page</title>

    <meta http-equiv="pragma" content="no-cache">
    <meta http-equiv="cache-control" content="no-cache">
    <meta http-equiv="expires" content="0">
    <meta http-equiv="keywords" content="keyword1,keyword2,keyword3">
    <meta http-equiv="description" content="This is my page">
    <!--
    <link rel="stylesheet" type="text/css" href="styles.css">
    -->

  </head>

  <body>
    <%--设置 username 属性，其值为 application_username --%>
    <jsp:setProperty name="user" property="username" value="application_username"/>
    <%--设置 password 属性，其值为 application_password --%>
    <jsp:setProperty name="user" property="password" value="application_password"/>
    <%--跳转到 getApplicationBean.jsp --%>
    <a href="getApplicationBean.jsp">
        跳转到 getApplicationBean.jsp</a>

  </body>
</html>
```

getApplicationBean.jsp

```jsp
<%@ page language="java" import="java.util.*" pageEncoding="utf-8"%>
<%--通过 useBean 动作指定调用 JavaBean --%>
<jsp:useBean id="user" scope="application" class="net.bean.User"></jsp:useBean>
<%
```

```jsp
    String path = request.getContextPath();
    String basePath = request.getScheme()+"://"+request.getServerName()+":"
        +request.getServerPort()+path+"/";
%>

<!DOCTYPE HTML PUBLIC "-//W3C//DTD HTML 4.01 Transitional//EN">
<html>
  <head>
    <base href="<%=basePath%>">

    <title>My JSP 'getApplicationBean.jsp' starting page</title>

    <meta http-equiv="pragma" content="no-cache">
    <meta http-equiv="cache-control" content="no-cache">
    <meta http-equiv="expires" content="0">
    <meta http-equiv="keywords" content="keyword1,keyword2,keyword3">
    <meta http-equiv="description" content="This is my page">
    <!--
    <link rel="stylesheet" type="text/css" href="styles.css">
    -->

  </head>

  <body>
    <%--获得username 属性--%>
    用户名：<jsp:getProperty name="user" property="username"/><br>
    <%--获得password 属性--%>
    密  码：<jsp:getProperty name="user" property="password"/>

  </body>
</html>
```

示例代码 ApplicationJavaBeanDemo.jsp 通过<jsp:useBean>指令调用名为 UserBean 的 JavaBean，并设置其实例化对象名为 user，其存储范围为 application 范围。通过<jsp:setProperty>设置 JavaBean 的 username 属性值为 application_username，设置 JavaBean 的 password 属性值为 application_password。通过<jsp:forward>设置页面跳转到 ApplicationJavaBeanDemo2.jsp。ApplicationJavaBeanDemo2.jsp 通过< jsp:getProperty>获得 JavaBean 的 username 属性和 password 属性。因为 JavaBean 的存储范围为 application，只要服务器不重启，就能够在任意页面中获得，就算是重新打开浏览器也是可以的，如图 6-9 所示。

图 6-9 设置 application 范围的 JavaBean

5. 移除 JavaBean

JavaBean 会根据其设置的范围来决定其生命周期，当生命周期结束后，JavaBean 将自动移除。不过设计者也可以手动地移除该 JavaBean，从而节省内存。

JavaBean 的移除因不同范围的 JavaBean 而不同，分别通过调用 pageContext、request、session、application 的 removeAttribute(String name)方法来移除 page 范围、request 范围、session 范围及 application 范围的 JavaBean。其中 name 属性设置为实例化对象名,必须和<jsp:useBean>中的 ID 属性保持一致。

例 6-12：移除 JavaBean 的示例。

removeBean.jsp

```jsp
<%@ page language="java" contentType="text/html;charset=gb2312"%>
<%--通过 userBean 动作指定调用 JavaBean --%>
<jsp:useBean id="user" class="com.javaweb.UserBean" scope="page"/>
<html>
<head>
<title>移除 JavaBean</title>
</head>
<body>
    <%--设置 username 属性，其值为 admin--%>
    <jsp:setProperty name="user" property="username" value="admin"/>
    <%--设置 password 属性，其值为 root --%>
    <jsp:setProperty name="user" property="password" value="root"/>
    <%
    //移除 page 范围 JavaBean
    pageContext.removeAttribute("user");
    %>
    <%--获得 username 属性--%>
    <jsp:getProperty name="user" property="username"/>
    <%--获得 password 属性--%>
    <jsp:getProperty name="user" property="password"/>
</body>
</html>
```

removeBean.jsp 通过<jsp:useBean>指令调用名为 UserBean 的 JavaBean，并设置其实例化对象名为 user，其存储范围为 page 范围。通过<jsp:setProperty>设置 JavaBean 的 username 属性值为 admin，设置 JavaBean 的 password 属性值为 root。通过 pageContext 属性的 removeAttribute()方法移除 page 范围 JavaBean。通过<jsp:getProperty>获得 JavaBean 的 username 属性和 password 属性。因为在获得属性之前已经移除了该 JavaBean，所以会抛出 java.lang.NullPointerException，即空指针异常。

6. JavaBean 综合

通过创建一个名为 Student 的 JavaBean，并在 JSP 中两次调用该 JavaBean，设置其 JavaBean 的属性，并获得其属性。

例 6-13：首先创建一个 Student 类，该类包含 3 个属性，分别为 name、sex、age。然后给这 3 个属性添加 setter 和 getter 方法，代码如下。

```java
package net.bean;

public class Student {
    private String name;
    private String sex;
    private int age;
    private String telephone;
    private String email;

    public String getName() {
        return name;
    }
    public void setName(String name) {
        this.name = name;
    }
    public String getSex() {
        return sex;
    }
    public void setSex(String sex) {
        this.sex = sex;
    }
    public int getAge() {
        return age;
    }
    public void setAge(int age) {
        this.age = age;
    }
    public String getTelphone() {
        return telephone;
    }
    public void setTelphone(String telephone) {
        this.telephone = telephone;
    }
    public String getEmail() {
        return email;
    }
    public void setEmail(String email) {
        this.email = email;
    }
}
```

**例6-14**：创建一个 JSP 页面，在该页面中通过<jsp:useBean>来调用 JavaBean，并实例化两个 Student 对象，分别为 student1 和 student2。然后在该 JSP 文件中设置并获得这两个对象的属性值。

```jsp
<%@ page language="java" import="java.util.*" pageEncoding="utf-8"%>
<%--通过 userBean 动作指定调用 StudentBean --%>
<jsp:useBean id="student1" scope="page" class="net.bean.Student"></jsp:useBean>
```

```jsp
<jsp:useBean id="student2" scope="page" class="net.bean.Student"></jsp:useBean>
<%
String path = request.getContextPath();
String basePath = request.getScheme()+"://"+request.getServerName()+":
   "+request.getServerPort()+path+"/";
%>

<!DOCTYPE HTML PUBLIC "-//W3C//DTD HTML 4.01 Transitional//EN">
<html>
  <head>
    <base href="<%=basePath%>">
    
    <title>My JSP 'studentBean.jsp' starting page</title>
    
    <meta http-equiv="pragma" content="no-cache">
    <meta http-equiv="cache-control" content="no-cache">
    <meta http-equiv="expires" content="0">
    <meta http-equiv="keywords" content="keyword1,keyword2,keyword3">
    <meta http-equiv="description" content="This is my page">
    <!--
    <link rel="stylesheet" type="text/css" href="styles.css">
    -->

  </head>
  
  <body>
    <%--设置 name 属性--%>
    <jsp:setProperty name="student1" property="name" value="王梅"/>
    <%--设置 sex 属性--%>
    <jsp:setProperty name="student1" property="sex" value="女"/>
    <%--设置 age 属性--%>
    <jsp:setProperty name="student1" property="age" value="21"/>
    <%--设置 telphone 属性--%>
    <jsp:setProperty name="student1" property="telphone" value="13888889999"/>
    <%--设置 email 属性--%>
    <jsp:setProperty name="student1" property="email" value="java999@163.com"/>
    
    <%--设置 name 属性--%>
    <jsp:setProperty name="student2" property="name" value="李明"/>
    <%--设置 sex 属性--%>
    <jsp:setProperty name="student2" property="sex" value="男"/>
    <%--设置 age 属性--%>
    <jsp:setProperty name="student2" property="age" value="27"/>
    <%--设置 telphone 属性--%>
    <jsp:setProperty name="student2" property="telphone" value="13988889999"/>
    <%--设置 email 属性--%>
    <jsp:setProperty name="student2" property="email" value="javafan@163.com"/>
    
    <table border="1">
        <tr><td>姓名</td> <td>性别</td><td>年龄</td> <td>电话</td><td>邮箱
```

```
                </td></tr>
            <tr>
                <td><jsp:getProperty name="student1" property="name"/></td>
                <td><jsp:getProperty name="student1" property="sex"/></td>
                <td><jsp:getProperty name="student1" property="age"/></td>
                <td><jsp:getProperty name="student1" property="telephone"/></td>
                <td><jsp:getProperty name="student1" property="email"/></td>
            </tr>
            <tr>
                <td><jsp:getProperty name="student2" property="name"/></td>
                <td><jsp:getProperty name="student2" property="sex"/></td>
                <td><jsp:getProperty name="student2" property="age"/></td>
                <td><jsp:getProperty name="student2" property="telephone"/></td>
                <td><jsp:getProperty name="student2" property="email"/></td>
            </tr>
        </table>
    </body>
</html>
```

运行结果如图 6-10 所示。

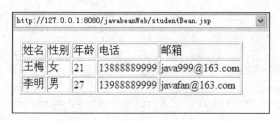

图 6-10 使用 JavaBean 的 JSP 应用

## 6.3 Java Web 开发模型

运用 JSP/Servlet 实现的 Web 动态交互，主要采用 JSP 和 JavaBean 模型还有 JSP 和 JavaBean+Servlet 模型。下面分别介绍这两种模型，并比较它们的优缺点。

### 6.3.1 JSP 和 JavaBean 开发模型

模型的结构如图 6-11 所示，称为 JSP 和 JavaBean 模型。其工作原理是：当浏览器发出请求时，JSP 接收请求并访问 JavaBean，若需要访问数据库或后台服务器，则通过 JavaBean 连接数据库或后台服务器，执行相应的处理。JavaBean 将处理的结果数据交给 JSP。JSP 提取结果并重新组织后，动态生成 HTML 页面，返回给浏览器。用户从浏览器显示的页面中得到交互的结果。

JSP 和 JavaBean 模型充分利用了 JSP 技术易于开发动态网页的特点，页面显示层的任务由 JSP（但它也含事物逻辑层的内容）承担，JavaBean 主要负责事务逻辑层和数据层的工作。JSP 和 JavaBean 模型依靠几个 JavaBean 组件实现具体的应用功能，生成动态内容，其最大的特点就是简单。

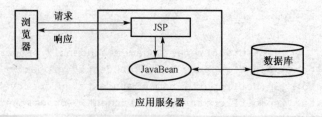

图 6-11　JSP 和 JavaBean 开发模型

**例 6-15：** 应用 JSP 和 JavaBean 模型的示例。用 JavaBean 实现业务功能的扩展，将业务功能放在 JavaBean 中完成，以使 JSP 页面程序更干净简洁，利于维护。JavaBean 可以很方便地用来捕获页面表单的输入，并完成各种业务逻辑的处理。

input.jsp

```jsp
<%@ page language="java" import="java.util.*" pageEncoding="gbk"%>
<%
String path = request.getContextPath();
String basePath = request.getScheme()+"://"+request.getServerName()+":"
    +request.getServerPort()+path+"/";
%>

<!DOCTYPE HTML PUBLIC "-//W3C//DTD HTML 4.01 Transitional//EN">
<html>
  <head>
    <base href="<%=basePath%>">

    <title>My JSP 'input.jsp' starting page</title>

    <meta http-equiv="pragma" content="no-cache">
    <meta http-equiv="cache-control" content="no-cache">
    <meta http-equiv="expires" content="0">
    <meta http-equiv="keywords" content="keyword1,keyword2,keyword3">
    <meta http-equiv="description" content="This is my page">
    <!--
    <link rel="stylesheet" type="text/css" href="styles.css">
    -->
<script language="javascript" type="text/javascript">
function f_check(){
    if(document.forms(0).username.value==""){
        alert("请输入姓名");
    }else{
        document.forms(0).submit();
    }
}
-->
</script>

  </head>
```

```html
  <body>
    <form name="form1" method="post" action="result.jsp" >
        <p>
        你的姓名：
        <input type="text" size="15" name="username" value="" id=username>
        <input type="button" align="center" name="tijiao" value="[提交]"
            onClick="f_check()" id=tijiao>
        </p>
    </form>
  </body>
</html>
```

result.jsp

```jsp
<%@ page language="java" import="java.util.*" pageEncoding="gbk"%>
<jsp:useBean id="serviceBean" scope="page" class="net.bean.ServiceBean">
<jsp:setProperty name="serviceBean" property="*" /></jsp:useBean>
<%
String path = request.getContextPath();
String basePath = request.getScheme()+"://"+request.getServerName()+":"
   +request.getServerPort()+path+"/";
%>

<!DOCTYPE HTML PUBLIC "-//W3C//DTD HTML 4.01 Transitional//EN">
<html>
  <head>
    <base href="<%=basePath%>">

    <title>My JSP 'result.jsp' starting page</title>

    <meta http-equiv="pragma" content="no-cache">
    <meta http-equiv="cache-control" content="no-cache">
    <meta http-equiv="expires" content="0">
    <meta http-equiv="keywords" content="keyword1,keyword2,keyword3">
    <meta http-equiv="description" content="This is my page">
    <!--
    <link rel="stylesheet" type="text/css" href="styles.css">
    -->

  </head>

  <body>
    <p>
    <%=serviceBean.hello()%>
    </p>
  </body>
</html>
```

TestBean.java

```java
package net.bean;

public class ServiceBean {
    private String username;

    public String hello(){
        String strHello = "Hello:"+this.getUsername()+"!";
        return strHello;
    }

    //汉字转换方法
    public String convertGBK(String str) {
        String strReturn = "";
        try {
            strReturn = new String(str.getBytes("ISO-8859-1"), "GBK");
        } catch (Exception ex) {
            System.out.println("TestBean.ConvertGBK():ex=" + ex.toString());
        } finally {
            return strReturn;
        }
    }

    public String getUsername() {
        return username;
    }

    public void setUsername(String username) {
        this.username = convertGBK(username);
    }
}
```

单击 input.jsp 页面上的"提交"按钮将表单提交给 result.jsp 页面，result.jsp 获得 input.jsp 中 username 文本框的值，并在实例化 ServiceBean 后，执行 bean 中的 setUsername 方法，接着执行 hello 方法，在页面上输出对用户问好的语句，如图 6-12、图 6-13 所示。

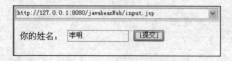 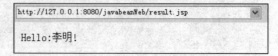

图 6-12  JSP+JavaBean 开发模型输入页面　　图 6-13  JSP+JavaBean 开发模型结果页面

这个简单的示例体现了在 JSP 中使用 JavaBean 的两个重要方面，一个是捕获表单的输入并保存，一个是执行逻辑功能。所以，依此两个功能还可以将用在 JSP 中的 JavaBean 分为值 Bean(value bean)和工具 Bean (utility bean)。

值 Bean 如下：

```java
package net.bean;
public class ValueBean{
    private String yourName = "";

    public void setUsername(String username){
        this.username = ConvertGBK(username);
    }

    //汉字转换方法
    public String ConvertGBK(String str){
        String strReturn="";
            try{
                strReturn=new String(str.getBytes("ISO-8859-1"),"GBK");
            }catch(Exception ex){
                System.out.println("ValueBean.ConvertGBK():ex="+ex.toString());
            }
            finally{
                return strReturn;
            }
    }
}
```

工具 Bean 如下：

```java
package net.bean;
public class UtilityBean{
    public String hello(ValueBean vBean){
        String strHello = "Hello:"+vBean.getUserame();
        return strHello;
    }

    public String hello(String username){
        String strHello = "Hello:"+ username;
        return strHello;
    }
}
```

当然，从这个例子看是没有必要分开值 Bean 和工具 Bean 的，但在具有复杂业务逻辑的 Web 应用程序中就可以用值 Bean 实现对表单输入的捕获、保存，减少对数据库中那些值几乎不变的实体的访问，或将值 Bean 放在一定作用域内使此作用域内的多个 JSP 页面共享。用工具 Bean 完成操作数据库、数据处理等业务逻辑，以值 Bean 或页面传递的值为参数。

### 6.3.2 JSP+JavaBean+Servlet 开发模型

JSP+JavaBean+Servlet 模型的体系结构如图 6-14 所示，称为 JSP 和 JavaBean 和 Servlet 模型。它是一种采用基于模型视图控制器（Model View Controller）的设计模型，即 MVC 模型。该模型将 JSP 程序的功能分为 3 个层次：Model（模型）层、View（视图）层、Controller（控制器层）。Model 层实现业务逻辑，包括了 Web 应用程序功能的核心，负责存储与应用程序相关的数据；View 层用于用户界面的显示，它可以访问 Model 层的数据，但不能更改这些数据；Controller 层主要负责 Model 和 View 层之间的控制关系。

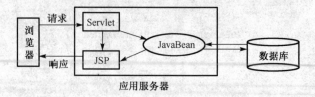

图 6-14　JSP+JavaBean+Servlet 开发模型

具体实现时，JavaBean 作为模型层，Servlet 作为控制层，JSP 作为视图层。每层的作用如下：

（1）JavaBean 作为 Model 层，实现各个具体的应用逻辑和功能。

（2）Servlet 作为 Controller 层，负责处理 HTTP 请求，包括：
- 对输入数据的检查和转换。
- 通过 JavaBean 访问数据库。
- 初始化 JSP 页面中要用到的 JavaBean 或对象。
- 根据处理中不同的分支和结果，决定转向哪个 JSP 等。

（3）JSP 作为 View 层，负责生成交互后返回的界面。它主要通过信息共享，获取 Servlet 生成的对象或 JavaBean，从中取出数据，插入到 HTML 页面中。

该模型的工作原理是：所有的请求都被发送给作为控制器的 Servlet。Servlet 接收请求，并根据请求信息将它们分发给相应的 JSP 页面来响应；同时 Servlet 还根据 JSP 的需求生成相应的 JavaBean 对象并传输给 JSP。JSP 通过直接调用方法或利用 UseBean 的自定义标签，得到 JavaBean 中的数据。

这种设计模式通过 Servlet 和 JavaBean 的合作来实现交互处理，很好地实现了表示层、事务逻辑层和数据的分离。

**例 6-16**：简单的用户登录过程示例。用户登录页面发起" *.do"的请求，被 web.xml 配置给 ControlServlet 进行处理，在 ControlServlet 中根据"*"的字符串（即解析用户请求的路径），调用 JavaBean 对象，在将处理后的 URL 转发给用户。

1. ControlServlet

```java
public class ControlServlet extends HttpServlet {
    public void doGet(HttpServletRequest request, HttpServletResponse
        response) throws ServletException, IOException {
        doPost(request, response);
    }

    public void doPost(HttpServletRequest request, HttpServletResponse
        response) throws ServletException, IOException {
        //得到用户名和密码
        String userName = request.getParameter("userName");
        String userPwd = request.getParameter("userPwd");
        if (checkUser(userName, userPwd)) {
            request.setAttribute("username", username);
```

```
                request.getRequestDispatcher("main.jsp").forward(request, response);
            } else {
                request.getRequestDispatcher("login.jsp").forward(request,
                    response);
            }
        }
    }
```

### 2. web.xml 配置

视图页面中的请求都是以<动作名字>.do 结尾，当这个请求到达 Web 服务器后，会被服务器转向给控制器处理，控制器在根据解析出的动作名，调用对应的 Action 对象，处理结果，并输出结果页面。所以 web.xml 中必须有如下配置：

```xml
<servlet>
    <servlet-name>controlServlet</servlet-name>
    <servlet-class>cn.netjava.servlet.ControlServlet</servlet-class>
    <init-param>
        <param-name>loginAction</param-name>
        <param-value>cn.netjava.action.LoginAction</param-value>
    </init-param>
</servlet>
<servlet-mapping>
    <servlet-name>controlServlet</servlet-name>
    <url-pattern>*.do</url-pattern>
</servlet-mapping>
```

### 3. JavaBean

```java
public class User
{
    private int userid;
    private int userclass;
    private String username;
    private String password;

    //业务方法
    public boolean checkUser(String username,String password){
        boolean flag;
        if (username.equals("admin") && password.equals("123")) {
            flag = true;
        } else {
            flag = false;
        }
        return flag;
    }

    public int getUserid() {
        return userid;
    }
```

```java
    public void setUserid(int userid) {
        this.userid = userid;
    }
    public int getUserclass() {
        return userclass;
    }
    public void setUserclass(int userclass) {
        this.userclass = userclass;
    }
    public String getUsername() {
        return username;
    }
    public void setUsername(String username) {
        this.username = username;
    }
    public String getPassword() {
        return password;
    }
    public void setPassword(String password) {
        this.password = password;
    }
}
```

4. View

如果登录成功，跳转到 main.jsp 页面，否则，返回 login.jsp 页面。

main.jsp

```jsp
<%@ page language="java" import="java.util.*" pageEncoding="utf-8"%>
<%
String path = request.getContextPath();
String basePath = request.getScheme()+"://"+request.getServerName()+":"
    +request.getServerPort()+path+"/";
%>

<!DOCTYPE HTML PUBLIC "-//W3C//DTD HTML 4.01 Transitional//EN">
<html>
  <head>
    <base href="<%=basePath%>">
    
    <title>My JSP 'main.jsp' starting page</title>
    
    <meta http-equiv="pragma" content="no-cache">
    <meta http-equiv="cache-control" content="no-cache">
    <meta http-equiv="expires" content="0">
    <meta http-equiv="keywords" content="keyword1,keyword2,keyword3">
    <meta http-equiv="description" content="This is my page">
    <!--
    <link rel="stylesheet" type="text/css" href="styles.css">
```

```
    -->
  </head>

  <body>
    <h1 style="color:red"><%=request.getAttribute("username") %>登录成功</h1>
  </body>
</html>
```

login.jsp

```
<%@ page language="java" import="java.util.*" pageEncoding="utf-8"%>
<%
String path = request.getContextPath();
String basePath = request.getScheme()+"://"+request.getServerName()+":"
    +request.getServerPort()+path+"/";
%>

<!DOCTYPE HTML PUBLIC "-//W3C//DTD HTML 4.01 Transitional//EN">
<html>
  <head>
    <base href="<%=basePath%>">

    <title>My JSP 'login.jsp' starting page</title>

    <meta http-equiv="pragma" content="no-cache">
    <meta http-equiv="cache-control" content="no-cache">
    <meta http-equiv="expires" content="0">
    <meta http-equiv="keywords" content="keyword1,keyword2,keyword3">
    <meta http-equiv="description" content="This is my page">
    <!--
    <link rel="stylesheet" type="text/css" href="styles.css">
    -->

  </head>

  <body>
    <form action="controlServlet" method="post">
    用户名：<input type="text" name="username" id="username"><br>
    密    码：<input type="password" name="userpwd"
        id="userpwd"><br>
    <input type="submit" value="登录"/>
    </form>

  </body>
</html>
```

运行结果如图 6-15、图 6-16 所示。

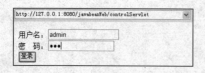

图 6-15　JSP+JavaBean+Servlet 开发模型登录页面　　图 6-16　JSP+JavaBean+Servlet 模型登录成功页面

### 6.3.3　两种开发模型比较

　　SP+JavaBean 模型和 JSP+JavaBean+Servlet 模型的整体结构都比较清晰，易于实现。它们的基本思想都是实现表示层、事务逻辑层和数据层的分离，这样的分层设计便于系统的维护和修改。两种模型的主要区别如下：

　　（1）处理流程的主控部分不同。JSP 和 JavaBean 模型利用 JSP 作为主控部分，将用户的请求、JavaBean 和响应有效地连接起来。JSP+JavaBean+Servlet 模型利用 Servlet 作为主控部分，将用户的请求、JavaBean 和响应有效地连接起来。

　　（2）实现表示层、事务逻辑层和数据层的分离程度不同。JSP+JavaBean+Servlet 模型比 JSP 和 JavaBean 模型有更好的分离效果。当事务逻辑比较复杂、分支较多或需要涉及多个 JavaBean 组件时，JSP 和 JavaBean 模型常会导致 JSP 文件中嵌入大量的脚本或 Java 代码。特别是大型项目开发中，由于页面设计和逻辑处理分别由不同的专业人员承担，如果 JSP 有相当一部分处理逻辑和页面描述混在一起，这就有可能引起分工不明确，不利于两个部分的独立开发和维护，影响项目的施工和管理。在 JSP+JavaBean+Servlet 模型中，由 Servlet 处理 HTTP 请求，JavaBean 承担事务逻辑处理，JSP 仅负责生成网页的工作，所以表现层的混合问题比较轻，适合于不同专业的专业人员独立开发 Web 项目中的各层功能。

　　（3）适应于动态交互处理的需求不同。当事务逻辑比较负责、分支较多或需要涉及很多 JavaBean 组件时，由于 JSP+JavaBean+Servlet 模型比 JSP 和 JavaBean 模型有更清晰的页面表现、更明确的开发模块的划分，所以使用 JSP+JavaBean+Servlet 模型比较适合。然而，JSP+JavaBean+Servlet 模型需要编写 Servlet 程序，Servlet 程序需要的工具是 Java 集成开发环境，编程工作量比较大。对于简单的交互处理，利用 JSP 和 JavaBean 模型，JSP 主要是使用 HTML 工具开发，然后再插入少量的 Java 代码就可以实现动态交互。在这种情况下，使用 JSP 和 JavaBean 模型更加方便快捷。

　　JSP 和 JavaBean 模型和 JSP+JavaBean+Servlet 模型这两种用于开发 Web 应用的方法都有很好的实用性。当然，实现动态交互的 Web 应用不限于这两种模型。在实际 Web 应用的开发过程中，需要根据系统特点、客户需求及处理逻辑的特性，选择合适的模型，力求使整个应用的体系结果更加合理，从而实现不同的交互处理。

## 6.4　分页 JavaBean 开发

　　鉴于 JSP 开发中经常用到分页显示的功能，故将分页显示进行统一设计和功能封装，在一定程度上可以加快开发进程，提高编码效率，减少重复性劳动，提高了代码重用度。

1. 页面基础类

```
public class Page {
    /**field*/
```

```java
private int currentPage = 1;            //当前页
private int index = 0;                  //当前页的索引
private int totalPage;                  //共有多少页
private int totalCount;                 //总的纪录数
private int everyPage = 10;             //每页显示的数据条数
private boolean hasNextPage;//是否有下一页,true 表示有,false 表示没有
private boolean hasPrePage;//是否有上一页,true 表示有,false 表示没有

/**default constructor*/
public Page() {}

/**full constructor*/
public Page(int everyPage, int totalCount, int totalPage, int currentPage,
        int index, boolean hasPrePage, boolean hasNextPage) {
    this.everyPage = everyPage;
    this.totalCount = totalCount;
    this.totalPage = totalPage;
    this.currentPage = currentPage;
    this.index = index;
    this.hasPrePage = hasPrePage;
    this.hasNextPage = hasNextPage;
}

/**get and set method*/
public int getCurrentPage() {
    return currentPage;
}
public void setCurrentPage(int currentPage) {
    this.currentPage = currentPage;
}
public int getIndex() {
    return index;
}
public void setIndex(int index) {
    this.index = index;
}
public int getTotalPage() {
    return totalPage;
}
public void setTotalPage(int totalPage) {
    this.totalPage = totalPage;
}
public int getEveryPage() {
    return everyPage;
}
public void setEveryPage(int everyPage) {
    this.everyPage = everyPage;
}
```

```java
    public boolean isHasNextPage() {
        return hasNextPage;
    }
    public void setHasNextPage(boolean hasNextPage) {
        this.hasNextPage = hasNextPage;
    }
    public boolean isHasPrePage() {
        return hasPrePage;
    }
    public void setHasPrePage(boolean hasPrePage) {
        this.hasPrePage = hasPrePage;
    }
    public int getTotalCount() {
        return totalCount;
    }
    public void setTotalCount(int totalCount) {
        this.totalCount = totalCount;
    }
}
```

2. 页面的执行类

```java
public class HandlePage {

    /**
     * 计算总共的页数
     *
     * @param count
     *            总共的纪录数
     * @param page
     *            当前被记录的页面对象
     * @return 返回计算的总页数
     */
    public static int getTotalPage(Page page) {
        int totalCount = page.getTotalCount();
        int totalPage = totalCount % page.getEveryPage() == 0 ? totalCount
                / page.getEveryPage() : totalCount / page.getEveryPage() + 1;
        return totalPage;
    }


    /**
     * 判断是否有下一页
     *
     * @param page
     *            当前的页面对象
     * @return 返回是否有下一页
     */
    public static boolean hasNextPage(Page page) {
        int totalPage = page.getTotalPage();
```

```java
        int currentPage = page.getCurrentPage();

        boolean hasNextPage = currentPage >= totalPage ? false : true;
        return hasNextPage;
    }

    /**
     * 判断是否有上一页
     *
     * @param page
     *            当前的页面对象
     * @return 返回是否有上一页
     */
    public static boolean hasPrePage(Page page) {
        int currentPage = page.getCurrentPage();

        boolean hasPrePage = (currentPage == 1) ? false : true;
        return hasPrePage;
    }

    /**
     * 返回当前页的索引
     *
     * @param page
     *            当前的页面对象
     * @return 返回当前页的索引
     */
    public static int getIndex(Page page) {
        return (page.getCurrentPage() - 1) * page.getEveryPage();
    }
}
```

3. 生成页面对象类

```java
public class ObtainPage {

    /**
     * 返回执行一次后的页面对象
     *
     * @param currentPage
     *            当前页
     * @param totalPage
     *            总共的页数
     * @param everyPage
     *            每页显示的纪录数
     * @param index
     *            当前页的索引
     * @param totalCount
     *            总共的纪录数
```

```
 * @param hasNextPage
 *            是否有下一页
 * @param hasPrePage
 *            是否有上一页
 * @param page
 *            执行页的当前页的对象
 * @return 返回当前页的对象
 */
public static Page getCurrentPage(int currentPage, int totalPage,
        int everyPage, int index, int totalCount, boolean hasNextPage,
        boolean hasPrePage, Page page) {
    page.setCurrentPage(currentPage);
    page.setEveryPage(everyPage);
    page.setTotalPage(totalPage);
    page.setTotalCount(totalCount);
    page.setIndex(index);
    page.setHasNextPage(hasNextPage);
    page.setHasPrePage(hasPrePage);
    return page;
}

/**
 * 返回执行一次后的页面对象
 *
 * @param currentPage
 *            当前页
 * @param index
 *            当前页的索引
 * @param hasNextPage
 *            是否有下一页
 * @param hasPrePage
 *            是否有上一页
 * @param page
 *            执行页的当前页的对象
 * @return 返回当前页的对象
 */
public static Page getCurrentPage(int currentPage, int index,
        boolean hasNextPage, boolean hasPrePage, Page page) {
    page.setCurrentPage(currentPage);
    page.setTotalCount(page.getTotalCount());
    page.setIndex(index);
    page.setHasNextPage(hasNextPage);
    page.setHasPrePage(hasPrePage);
    page.setEveryPage(page.getEveryPage());
    page.setTotalPage(HandlePage.getTotalPage(page));
    return page;
}
```

```java
    /**
     * 返回执行一次后的页面对象
     *
     * @param totalCount
     *            总共的纪录数
     * @param page
     *            执行页的当前页的对象
     * @return 返回当前页的对象
     */
    public static Page getCurretnPage(int totalCount, Page page) {
        page.setTotalCount(totalCount);
        page.setTotalPage(HandlePage.getTotalPage(page));
        page.setCurrentPage(page.getCurrentPage());
        page.setEveryPage(page.getEveryPage());
        page.setHasNextPage(HandlePage.hasNextPage(page));
        page.setHasPrePage(HandlePage.hasPrePage(page));
        page.setIndex(HandlePage.getIndex(page));
        return page;
    }

    /**
     * 返回执行一次后的页面对象
     *
     * @param currentPage
     *            当前页
     * @param totalCount
     *            总共的纪录数
     * @param page
     *            当前页的对象
     * @return 返回当前页的对象
     */
    public static Page getCurretnPage(int currentPage, int totalCount, Page page){
        page.setTotalCount(totalCount);
        page.setTotalPage(HandlePage.getTotalPage(page));
        page.setCurrentPage(currentPage);
        page.setEveryPage(page.getEveryPage());
        page.setHasNextPage(HandlePage.hasNextPage(page));
        page.setHasPrePage(HandlePage.hasPrePage(page));
        page.setIndex(HandlePage.getIndex(page));
        return page;
    }
}
```

4. JSP 页面使用分页 JavaBean

```
<%@ page language="java" import="net.bbs.util.PageBean,net.bbs.bean.*,
                java.util.*" pageEncoding="UTF-8"%>
<!DOCTYPE HTML PUBLIC "-//W3C//DTD HTML 4.01 Transitional//EN">
<html>
```

```
    <head>
        <title>bbslist</title>
        <meta http-equiv="pragma" content="no-cache">
        <meta http-equiv="cache-control" content="no-cache">
        <meta http-equiv="expires" content="0">
        <meta http-equiv="keywords" content="keyword1,keyword2,keyword3">
        <meta http-equiv="description" content="This is my page">
        <!--
        <link rel="stylesheet" type="text/css" href="styles.css">
        -->
        <base target="rtop">
<link rel="stylesheet" type="text/css" href="bbs01.css">
<style type="text/css">
<!--
.header {font: 11px Tahoma, Verdana; color: #000000; font-weight: bold;
         background-color: #99CC66 }
.maintable {width: 100%; background-color: #EFFFCE }
.tableborder {background: #D6E0EF; border: 1px solid #000000 }
-->
</style>
<%
    PageBean pb = (PageBean) session.getAttribute("pb");

    int bbsid, parentid, child, bbshits, length;
    String useremail, userip, expression, usersign, bbstopic, bbscontent,
        bbshot;
    java.util.Date dateandtime = new java.util.Date();

    String username = (String) session.getAttribute("username");
    int boardid = pb.getBoard().getBoardid();
    String boardname = pb.getBoard().getBoardname();
    String boardmaster = pb.getBoard().getBoardmaster();
    String masterword = pb.getBoard().getMasterword();
    String masteremail = pb.getBoard().getMasteremail();
    int method = pb.getMethod();
    String delon = "false";
    if (method == 8) { // '管理员开关
        delon = "false";
        if (session.getAttribute("superlogin") != "true") {
            // '非正常版主登录
%>
<script language="JavaScript">
    window.location = "superlog.jsp";
</script>
<%
    } else {
        delon = "true";
    }//' end if
```

```html
        } //'End if
%>
</head>
<body>


<div align="center" bgcolor="#EFFFCE">
  <table cellspacing="1" cellpadding="4" width="757" align="center"
                bgcolor="#EFFFCE">
    <tr class="header"><td height="58"><h1><div align="center">BBS</h1>
          </td></tr>
  </table>
</div>

<div align="center">
<table border="0" width="757">
  <tr>
    <td align=middle width="18%"><a href="mailto:<%=masteremail%>">
          (<img src="images/online_moderator.gif"width="16" height="15"
          border="0"> 版主:<%=boardmaster%></a>)</td>
    <td width="17%" align=middle>主题: <b><i><%=boardname%></i></b></td>
    <td width="20%" align=middle><b><i><a href="loginServlet">返回主页
          </a></i></b></td>
    <td width="45%" align=middle>板块说明: <font color="#FF0000"><%=masterword%>
          </font></td>
  </tr>
</table>

<!--表头部分 --><!--表头部分结束 -->
<table border="1" width="759" bordercolorlight="#000000" bordercolordark=
          "#FFFFFF" bgcolor="#EFFFCE" cellspacing="0" cellpadding="3">
  <tr>
    <td width="40" bgcolor="#99CC66"><p align="center"><a href="bbsadd.jsp?
          boardid=<%=boardid%>&boardname=<%=boardname%>" target=
          "_self"><font color="#FFFFFF">留言</font></a></p></td>
    <td width="40" bgcolor="#99CC66"><p align="center"><a href="preBbslistServlet?
          boardid=<%=boardid%>" target="_self"><font color="#FFFFFF">
          刷新</font></a></p></td>
    <td width="40" bgcolor="#99CC66"><p align="center"><a href="query.jsp">
          <font color="#FFFFFF">查询</font></a></p></td>
    <td width="70" bgcolor="#99CC66"><p align="center">
          <a href="preBbslistServlet?boardid=<%=boardid%>
          &method=4"><font color="#FFFFFF">精华区</font></a></p></td>
    <td width="200" bgcolor="#99CC66"><p align="center">
      <%if (pb.getPageCount()>0) {%>
      <a href="preBbslistServlet?boardid=<%=boardid%>&pages=1">
          <font color="#FFFFFF">首页</font></a>
    <%}%>
```

```html
        <%if (pb.getCurrentPag()>1) {%>
        <a href="preBbslistServlet?boardid=<%=boardid%>&pages=<%=
               pb.getPagePre()%>"><font color="#FFFFFF">上页</font></a>
        <%}%>
        <%if (pb.getCurrentPag()<pb.getPageCount()) {%>
        <a href="preBbslistServlet?boardid=<%=boardid%>&pages=<%=
               pb.getPageNext()%>"><font color="#FFFFFF">下页</font></a>
        <%}%>
        <%if (pb.getPageCount()>1) {%>
        <a href="preBbslistServlet?boardid=<%=boardid%>&pages=<%=
               pb.getPageCount()%>"><font color="#FFFFFF">尾页</font></a>
        <%}%>
        <font color="#FFFFFF"><%=pb.getCurrentPag()%>/<%=pb.get-
               PageCount()%></font></p>
     </td>
     <td width="110" valign="middle" bgcolor="#99CC66">
      <form method="GET" action="preBbslistServlet" style=
              "margin-top: 0; margin-bottom: 0">
         <p align="center" style="margin-top: 0; margin-bottom: 0">
               <font color="#FFFFFF">
         转到:<input type="text" name="pages" size="3" value="1">页
         <input type="submit" value="GO" name="GO"></font></p>
         <input type="hidden" name="boardid" value="<%=boardid%>">
      </form>
     </td>
  </tr>
</table>
</div>

<div align="center">

  <table bgcolor="#EFFFCE" border="2" cellSpacing=1 cellPadding=4 width="761">
   <TR class=category>
    <td align=middle width="7%"> </td>
    <td align=middle width="23%">标题 </td>
    <td noWrap align=middle width="8%">大小</td>
    <td align=middle width="12%">作者</td>
    <td noWrap align=middle width="14%">发表日期</td>
    <td noWrap align=middle width="13%">作者ID</td>
    <td noWrap align=middle width="11%">点击率</td>
    <td align=middle width="11%">回复次数</td>
   </TR>
<%
List list = pb.getRowList();
if(list!=null&&list.size()>0){
//'如果是首次显示,则当前页为1,否则根据请求的页数显示
for(int j=0;j<list.size();j++){
    Bbs bbs = (Bbs)list.get(j);
```

```jsp
        bbsid=bbs.getBbsid();
        parentid=bbs.getParentid();
        child=bbs.getChild();

        username=bbs.getUsername();
        useremail=bbs.getUseremail();
        userip=bbs.getUserip();
        usersign=bbs.getUsersign();

        expression=bbs.getExpression();
        bbstopic=bbs.getBbstopic();
        bbscontent=bbs.getBbscontent();
        //dateandtime=sqlRst.getString("dateandtime");
        dateandtime = bbs.getDateandtime();

        bbshits=bbs.getBbshits();
        length=bbs.getBbslength();

        bbshot=bbs.getBbshot();
%>
<tr>
<td align=middle > <p align="left"><img src="images/06.gif" width=
        "45" height="45" border="0"></td>
<td align=middle ><a href="preBbsaddreServlet?boardid=<%=boardid%>
        &bbsid=<%=bbsid%>"><%=bbstopic%></a><img src=
        "images/agree.gif"></td>
<td align=middle > <font color="#FF0000"><%if (length==0)
        {%>无内容<%}else{%><%=length%>Bytes<%}%></font></td>
<td align=middle >【<a href="mailto:<%=useremail%>"><b><%=
        username%></b></a>】</td>
<td align=middle ><i><%=dateandtime%></i></td>
<td align=middle >ID:<%=bbsid%></td>
<td align=middle >点击:<%=bbshits%></td>
<td align=middle >回复:<%=child%></td>
</tr>
 <%if (bbshot.compareTo("ok")==0){%><font color="#FF0000">★</font><%}%>
 <%if (delon=="true"){%>
    <a href="delete.jsp?bbsid=<%=bbsid%>">删除</a>
   ###<%if(bbshot.compareTo("ok")!=0) {%><a href="addfavServlet?bbsid=<%=
        bbsid%>">加入精华区</a>
 <%   }else{%><a href="subfav.jsp?bbsid=<%=bbsid%>">从精华区删除</a>
     <%}
   }%>


<!--跟贴开始---------------------->

<!--跟贴结束---------------------->
```

```
<%
}//for
}else{//if
%>
<table  bgcolor="#EFFFCE" cellspacing="1" cellpadding="4" width=
                "761" align="center" class="tableborder">
<tr><td  align=middle height="32" colspan="7" class="header">
 系统中还没有记录!
<%
}%>
 </table>
</div>
<!--表头部分 -->
 <hr size="0" color="#808080">
<!--表头部分结束 -->
</body>
</html>
```

## 6.5 小  结

本章介绍了 JavaBeans 组件的概念和 JavaBeans 的编写和使用。通过本章的学习，应该了解 JavaBeans 编程约定，学会编写简单的 JavaBeans，掌握 JavaBeans 的配置，掌握应用 JSP 标记来使用 JavaBeans，掌握应用 JSP 和 JavaBean 的两种开发模型。

## 6.6 习  题

1. 编写一个用于记录论坛的用户信息的 JavaBean 类，类名取为 users，类中包括姓名、性别、住址、Email、工作、兴趣等属性。

2. 应用 JSP 标记来使用 users，设置和获取其属性。

3. 设置一个 HTML 表单 register.html，用于论坛的用户注册，提交的内容至少应包括：姓名、性别、住址、Email、工作、兴趣等。

4. 编写用于处理论坛的用户注册的 JSP 页面，使用 JavaBean users 来获取页面注册信息。

# 第 7 章  JSP 中的文件操作

本章主要讲解 Web 页面与文件之间的数据传输。掌握使用 File 类操作文件属性、文件和目录，掌握文件上传和下载的实现方法和技术。

## 7.1  File 类和数据流

### 7.1.1  File 类

File 类的对象主要用来获取文件本身的一些信息。例如文件所在的目录、文件的长度和文件读写权限等，不涉及对文件的读写操作。实际上一个 File 类的对象就代表一个文件（或目录）。

1. File 类的构造方法

- File(String filename)：filename 为文件名，该文件与当前应用程序在同一目录中。
- File(String directoryPath,String filename)：directoryPath 是文件路径。
- File(File file,String filename)：file 是指定成目录的一个文件。

2. File 类获取文件属性信息的方法

- public String getName()：获取文件的名字。
- public Boolean canRead()：判断文件是否可读。
- public Boolean canWrite()：判断文件是否可写。
- public Boolean exits()：判断文件是否存在。
- public long length()：获取文件的长度（单位是字节）。
- public String getAbsolutePath()：获取文件的绝对路径。
- public String getParent()：获取文件的父目录。
- public Boolean isFile()：判断文件是一个正常文件，而不是目录。
- public Boolean isDirectory()：判断文件是否是一个目录。
- public Boolean isHidden()：判断文件是否是隐藏文件。
- public long lastModified()：获取文件最后修改的时间。

**例 7-1**：使用 File 类获取文件属性方法，获取某些文件的信息。在 root 目录下创建一个文件 new.txt，然后测试该文件的属性信息。

```
<%@ page contentType="text/html;charset=utf-8" %>
<%@ page import="java.io.*"%>
<HTML>
<BODY bgcolor=cyan><Font Size=1>
  <%
File f1=new
```

```
        File("E:\\Tomcat\\webapps\\root","file.txt");
        F1.creatNewFile();
     %>
    <P> 文件 file.txt 存在吗?
       <%=f1.exists()%>
    <BR>
    <P>文件 file.txt 的长度:
       <%=f1.length()%>字节
    <BR>
    <P> 文件 file.txt 是可读的吗?
       <%=f2.canRead()%>
    <BR>
    <P> file.txt 的父目录是:
       <%=f1.getParent()%>
    </Font>
    </BODY>
    </HTML>
```

3. File 类的目录管理方法

（1）创建目录

```
    public boolean mkdir();   //创建目录
```

（2）列出目录中的文件

如果 File 对象是一个目录，那么该对象可以调用下述方法列出该目录下的文件和子目录：
- public String[] list()：用字符串形式返回。
- public File[] listFiles()：用 File 对象形式返回。

（3）列出指定类型的文件
- public String[] list(FilenameFilter obj)：字符串形式目录下指定类型的所有文件
- public File[] listFiles(FilenameFilter obj)：用 File 对象形式返回目录下指定类型的所有文件。其中 FilenameFilter 是一个接口，该接口有如下方法：

```
    public boolean accept(File dir,String name)
```

当向 list 方法传递一个实现该接口的对象时，dir 调用 list 方法在列出文件时，将调用 accept 方法检查该文件 name 是否符合 accept 方法指定的目录和文件名字要求。

（4）删除文件或目录

```
    public Boolean delete()
```

该方法可以删除当前对象代表的文件或目录，如果 File 对象表示的是一个目录，则该目录必须为空，删除成功返回 true。

例 7-2：删除 root/files 目录下的文件 file.txt，然后删除目录 files。

```
    <%@ page contentType="text/html;charset=utf-8" %>
    <%@ page import ="java.io.*" %>
    <HTML>
    <BODY>
```

```
    <%
File f=new File("E:\\Tomcat\\webapps\\root\\files","file.txt");
File dir=new File("E:\\Tomcat\\webapps\\root\\files");
    boolean b1=f.delete();
    boolean b2=dir.delete();
  %>
<P>文件 new.doc 成功删除了吗?
    <%=b1%>
<P>目录 Students 成功删除了吗?
    <%=b2%>
</BODY>
</HTML>
```

例 7-3：File 类简单用法。

```
public class TestFile {
    public void createFile(String path){
        File file=new File(path);
        if(!file.exists()){                    //判断文件是否存在
            try {
                file.createNewFile();          //创建文件
            } catch (IOException e) {
                e.printStackTrace();
            }
        }
        /*获取文件名*/
        String name=file.getName();
        /*获取文件路径*/
        String path_=file.getPath();
        /*获取绝对路径名*/
        String absPath=file.getAbsolutePath();
        /*获取父亲文件路径*/
        String parent=file.getParent();
        /*文件大小*/
        long size=file.length();
        /*最后一次修改时间*/
        long time=file.lastModified();
        System.out.println("文件名:"+name);
        System.out.println("文件路径:"+path_);
        System.out.println("文件的绝对路径:"+absPath);
        System.out.println("文件的父文件路径:"+parent);
        System.out.println("文件的大小:"+size);
        System.out.println("文件最后一次修改时间:"+time);
        //file.delete();    删除文件
    }

    public void createDir(String path){
        File file=new File(path);
        if(!file.exists()){
```

```java
            file.mkdirs();   //创建文件夹
        }
        
        //file.delete();   若文件夹为空,则删除文件夹
    }
    
    /**
     * 遍历文件夹中的文件并显示
     */
    public void fileTest(String path){
        File file=new File(path);
        File[] files=file.listFiles();
        
        for (File f : files) {
            if(f.isFile()){
                System.out.println(f.getName()+"是文件!");
            }else if(f.isDirectory()){
                fileTest(f.getPath());
            }
        }
    }
    
    public void reFileName(String fromPath,String toPath){
        File file1=new File(fromPath);
        File file2=new File(toPath);
        /*判断file2文件夹路径存在与否,不存在则创建*/
        if(!file2.exists()){
            new File(file2.getParent()).mkdirs();
        }
        file1.renameTo(file2);    //修改文件名
    }
    
    public static void main(String[] args) {
        //File file=new File("E:\\myjava\\1.txt");   // Window路径\\
        //File flie_=new File("E:/myjava","1.txt");  //linux路径 /
        
        TestFile tf=new TestFile();
        //tf.createFile("E:\\myjava\\1.txt");
        //tf.fileTest("E:/wepull");
        //tf.createDir("e:/sunxiao/abc/1.txt");
        //tf.reFileName("E:\\my\\2.txt","E:\\myjava\\1.txt");
        tf.copyFlie("E:\\myjava","F:\\");
    }
    
    private void copyFlie(String src, String to) {
        File file1=new File(src);
        String topath=to+"\\"+file1.getName();
        File file2=new File(topath);
```

```
            if(!file2.exists()){
               file2.mkdirs();
            }
         System.out.println(topath);
         File[] file=file1.listFiles();
         for (File f : file) {
             if(f.isFile()){
                 String path2=topath+"\\"+f.getName();
                 Creatfile(path2);
             }else if(f.isDirectory()){
                 String s=f.getPath();
                 copyFlie(s,topath);
             }
         }
    }

    private void Creatfile(String path2) {
        File file3=new File(path2);
        if(!file3.exists()){//判断文件是否存在
            try {
                file3.createNewFile();  //创建文件
            } catch (IOException e) {
                e.printStackTrace();
            }
        }
    }
}
```

### 7.1.2 字节输入流类和字节输出流类

所有字节输入流类都是 InputStream 抽象类的子类，而所有的字节输出流类都是 OutputStream 抽象类的子类。FileInputStream、FileOutputStream 类分别从 InputStream、OutputStream 类继承而来。

#### 1. FileInputStream

FileInputStream 是从 InputStream 中派生出来的简单输入类，该类提供了基本的文件写入功能，顺序写。

（1）构造方法
- FileInputStream(String name)：使用给定的文件名创建对象。
- FileInputStream(File file)：使用 File 对象创建 FileInputStream 对象。

（2）读取文件

使用文件输入流读取文件：

```
FileInuputStream istream=new FileInputStream("myfile.doc");
```

或：

```
File f=new File("myfile.doc");
FileInputStream istream=new FileInputStream(f);
```

构造方法可能会产生异常IOException,为了把一个文件输入流对象与一个文件关联起来,需要try和catch语句。

```
try {
    FileInputStream ins=new FileInputStream("myfile.doc");
}catch (IOException e){
    System.out.println("File read error:"+e);
}
```

(3) 从输入流中读取字节
- int read()：从输入流中读取单个字节数据（0～255），如到输入流末尾则返回-1。
- int read(byte b[])：读多个字节。
- int read(byte b[],int off,int len)：从off开始读len个字节到输入流。

(4) 关闭流
- close()：关闭流。

Java在程序结束时自动关闭所有打开的流,但显式关闭任何打开的流是一个好习惯。

2. FileOutputStream

FileOutputStream是从OutputStream中派生出来的简单输出类,该类提供了基本的文件输出功能。

(1) 构造方法
- FileOutputStream(String name)：用给定的文件名创建对象。
- FileOutputStream(File file)：用File类创建对象。

(2) 输出方法
- public void write(byte b[])：写b.length个字节到输出流。
- public void write(byte b[],int off,int len)：从off开始写len个字节到输出流。

3. BufferedInputStream 类和 BufferedOutputStream 类

为了提高读写效率,FileInputStream经常和BufferedInputStream流配合使用,FileOuputStream流经常和BufferedOutputStream流配合使用。

对应的构造方法分别如下：
- BufferedInputStream(InputStream in)
- BufferedOutputStream(OutputStream out)

例7-4：BufferedInputStream类简单用法。

```
<%
    File f = new File("F:/test.java");
    try{
        FileOutputStream outfile = new FileOutputStream(f);
        BufferedOutputStream bufferout = new BufferedOutputStream(outfile);
        byte[] b = "nice to meet you ".getBytes();
        bufferout.write(b);
        bufferout.flush();
        bufferout.close();
```

```
        outfile.close();
        FileInputStream in = new FileInputStream(f);
        BufferedInputStream bufferin = new BufferedInputStream(in);
        byte c[] = new byte[90];
        int n = 0;
        while((n=bufferin.read(c))!=-1){
            String temp = new String(c,0,n);
            out.print(temp);
        }
        bufferin.close();
        in.close();
    }catch(IOException e){
        e.printStackTrace();
    }
%>
```

## 7.2 文件上传

### 7.2.1 JSP 页面处理文件上传

文件上传是 Web 应用经常需要面对的问题，在大部分时候，用户的请求参数是在表单域输入的字符串，但如果为表单元素设置 enctype="multipart/form-data"属性，则提交表单时不再以字符串方式提交请求参数，而是以二进制编码的方式提交请求，此时直接通过 HttpServletRequest 的 getParameter 方法无法正常获取请求参数的值，但可以通过二进制流来获取请求内容，通过这种方式，就可以取得希望上传文件的内容，从而实现文件的上传。

在一个 Web 工程中，要实现文件的上传，首先要把表单上传数据的编码方式设置为二进制数据方式，这就用到表单的 enctype 属性。这个属性可以有以下 3 个值：

（1）application/x-www-form-urlencoded：这是默认值，代表的方式适用范围比较广泛，只要是能输出页面的服务器环境都可以用。它处理表单域里的 Value 属性值，采用这种编码方式的表单会将表单域的值处理成 URL 编码方式。然而，在向服务器发送大量的文本、包含非 ASCII 字符的文本或二进制数据时，这种编码方式效率很低。

（2）multipart/form-data：上传二进制数据，这种编码方式会把文件域指定文件的内容也封装到请求参数中，只有使用这种编码方式，才能完整地传递文件数据，进行上传的操作。

（3）text/plain：这种编码方式当表单的 action 属性为 mailto:URL 的形式时比较方便，主要适用于直接通过表单发送电子邮件的方式。

客户通过 Web 页面上传文件给服务器的 File 类型表单如下：

```
<form action="接受上传文件的页面"method="post" ENCTYPE=
                                "multipart/form-data">
<input type="file" name="picture">
</form>
```

JSP 引擎可以在接受上传文件的页面中，通过内置对象 request 调用 getInputStream 方法获得一个输入流，通过这个输入流读入客户上传的全部信息，包括文件的内容以及表单域的信息。

一般地，文件表单提交的信息中，前4行和最后面的5行是表单本身的信息，中间部分才是客户提交的文件的内容。为此，我们要通过输入、输出流技术把表单的信息去掉，只获取文件的内容。

首先将客户提交的全部信息保存为一个临时文件；然后读取该文件的第2行，获取文件名字；再获取第4行的结束位置以及倒数第6行的结束位置，这两个位置之间的内容就是上传文件的内容；然后将这部分内容存入文件，文件名与客户上传时的文件名一样；最后删除临时文件。

**例7-5**：文件上传示例。

```jsp
fileJsp.jsp
<%@ page contentType="text/html;charset=utf-8" %>
<HTML>
<BODY>
        <P>选择要上传的文件：<BR>
        <form action="acceptFile.jsp" method="post" ENCTYPE="multipart/form-data">
        <%-- 类型enctype用multipart/form-data，这样可以把文件中的数据作为流
             据上传，不管是什么文件类型，均可上传--%>
        <input type="file" name="upfile" size="20">
        <BR>
        <input type="submit" value="提交">
    </form>
 </BODY>
</HTML>


acceptFile.jsp:
<%@ page contentType="text/html;charset=utf-8" %>
<%@ page import ="java.io.*" %>
<HTML>
<BODY>
  <%
    try {
        InputStream in = request.getInputStream();
        File f = new File("F:/files", "upfile.txt");
        FileOutputStream o = new FileOutputStream(f);
        byte b[] = new byte[1000];
        int n;
        while ((n = in.read(b)) != -1) {
            o.write(b, 0, n);
        }
        o.close();
        in.close();
    } catch (IOException e) {
        e.printStackTrace();
    }
    out.print("文件已上传");
  %>
</BODY>
</HTML>
```

运行结果如图 7-1 所示。

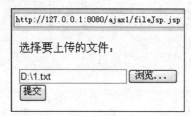 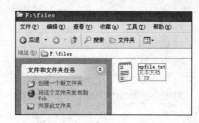

图 7-1　文件上传页面和文件保存页面

### 7.2.2　Servlet 处理文件上传

**例 7-6**：使用 Servlet 处理文件的上传。
upload.jsp

```
<form action="/ajax1/servlet/uploadServlet" method="post" ENCTYPE=
           "multipart/form-data">
    <input type="file" name="upfile" size="20">
    <BR>
    <input type="submit" value="提交">
  </form>
UploadServlet 类:
public class UploadServlet extends HttpServlet {
    @Override
    protected void doGet(HttpServletRequest req, HttpServletResponse resp)
            throws ServletException, IOException {
        doPost(req, resp);
    }

    @Override
    protected void doPost(HttpServletRequest request, HttpServletResponse
        response)
            throws ServletException, IOException {
        response.setContentType("text/html;charset=utf-8");
        request.setCharacterEncoding("utf-8");
        PrintWriter out = response.getWriter();
        //定义上载文件的最大字节
int MAX_SIZE = 102400 * 102400;
//创建根路径的保存变量
String rootPath;
//声明文件读入类
DataInputStream in = null;
FileOutputStream fileOut = null;
//取得客户端的网络地址
String remoteAddr = request.getRemoteAddr();
//获得服务器的名字
String serverName = request.getServerName();
//取得站点的绝对地址
```

```java
        String realPath = request.getSession().get-
                ServletContext().getRealPath("/");
realPath = realPath.substring(0, realPath.lastIndexOf("\\"));
//创建文件的保存目录
rootPath = realPath + "\\WebRoot\\upload\\";
//取得客户端上传的数据类型
String contentType = request.getContentType();
try {
    if (contentType.indexOf("multipart/form-data") >= 0) {
        //读入上传的数据
        in = new DataInputStream(request.getInputStream());
        int formDataLength = request.getContentLength();
        if (formDataLength > MAX_SIZE) {
            out.println("<P>上传的文件字节数不能超过" + MAX_SIZE + "</p>");
            return;
        }
        //保存上传文件的数据
        byte dataBytes[] = new byte[formDataLength];
        int byteRead = 0;
        int totalBytesRead = 0;
        //上传的数据保存在 byte 数组
        while (totalBytesRead < formDataLength) {
            byteRead = in.read(dataBytes, totalBytesRead,
                    formDataLength);
            totalBytesRead += byteRead;
        }
        //根据 byte 数组创建字符串
        String file = new String(dataBytes);
        //out.println(file);
        //取得上传的数据的文件名
        String saveFile = file.substring(file
                .indexOf("filename=\"") + 10);
        saveFile = saveFile.substring(0, saveFile.indexOf(
                "\n"));
        saveFile = saveFile.substring(
                saveFile.lastIndexOf("\\") + 1,
                saveFile.indexOf("\""));
        int lastIndex = contentType.lastIndexOf("=");
        //取得数据的分隔字符串
        String boundary = contentType.substring(lastIndex + 1,
                contentType.length());
        //创建保存路径的文件名
        String fileName = rootPath + saveFile;
        out.print(fileName);
        int pos;
        pos = file.indexOf("filename=\"");
        pos = file.indexOf("\n", pos) + 1;
        pos = file.indexOf("\n", pos) + 1;
```

```java
            pos = file.indexOf("\n", pos) + 1;
            int boundaryLocation = file.indexOf(boundary, pos) - 4;
            //out.println(boundaryLocation);
            //取得文件数据的开始的位置
            int startPos = ((file.substring(0, pos)).getBytes()).
                            length;
            //out.println(startPos);
            //取得文件数据的结束的位置
            int endPos = ((file.substring(0, boundaryLocation))
                    .getBytes()).length;
            //out.println(endPos);
            //检查上载文件是否存在
            File checkFile = new File(fileName);
            if (checkFile.exists()) {
                out.println("<p>" + saveFile + "文件已经存在.</p>");
            }
            //检查上载文件的目录是否存在
            File fileDir = new File(rootPath);
            if (!fileDir.exists()) {
                fileDir.mkdirs();
            }
            //创建文件的写出类
            fileOut = new FileOutputStream(fileName);
            //保存文件的数据
            fileOut.write(dataBytes, startPos, (endPos - startPos));
            fileOut.close();
            out.println(saveFile + "文件成功上载.</p>");
        } else {
            String content = request.getContentType();
            out.println("<p>上传的数据类型不是multipart/form-data
                        </p>");
        }
    } catch (Exception ex) {
        throw new ServletException(ex.getMessage());
    }
  }
}
```

web.xml

```xml
<servlet>
    <servlet-name>uploadServlet</servlet-name>
    <servlet-class>servlet.UploadServlet</servlet-class>
</servlet>

<servlet-mapping>
    <servlet-name>uploadServlet</servlet-name>
    <url-pattern>/servlet/uploadServlet</url-pattern>
</servlet-mapping>
```

## 7.3 文 件 下 载

JSP 内置对象 response 调用 getOutputStream（）方法可以获取一个指向客户的输出流，服务器将文件写入这个流，客户就可以下载这个文件了。

当 JSP 页面提供下载功能时，应当使用 response 对象向客户发送 HTTP 头信息，说明文件的 MIME 类型，这样客户的浏览器就会调用相应的外部程序打开下载的文件。

**例 7-7**：客户单击页面中的超链接时下载一个 zip 文档。

```jsp
<%@ page contentType="text/html;charset=utf-8" %>
<HTML>
<BODY>
<P>单击超链接下载 zip 文档 book.zip
<BR>  <A href="loadFile.jsp">下载 book.zip
</Body>
</HTML>
```

loadFile.jsp:
```jsp
<%@ page contentType="text/html;charset= utf-8" %>
<%@ page import="java.io.*" %>
<HTML>
<BODY>
<% //获得响应客户的输出流
   OutputStream o=response.getOutputStream();
   //输出文件用的字节数组,每次发送 500 个字节到输出流
   byte b[]=new byte[500];
   //下载的文件
   File fileLoad=new File("f:/2000","book.zip");
   //客户使用保存文件的对话框
   response.setHeader("Content-disposition","attachment;filename=
                      "+"book.zip");
   //通知客户文件的 MIME 类型
   response.setContentType("application/x-tar");
   //通知客户文件的长度
   long fileLength=fileLoad.length();
   String length=String.valueOf(fileLength);
   response.setHeader("Content_Length",length);
   //读取文件 book.zip,并发送给客户下载
   FileInputStream in=new FileInputStream(fileLoad);
   int n=0;
   while((n=in.read(b))!=-1) {
o.write(b,0,n);
   }
%>
</BODY>
</HTML>
```

JSP中实现文件下载的最简单的方式是在网页上做超级链接，如：<a href= "upload/file.doc">点击下载</a>。但是这样服务器上的目录资源会直接暴露给最终用户，会给网站带来一些不安全的因素。因此可以采用其他方式实现下载，可以采用 RequestDispatcher 的方式进行或采用文件流输出的方式下载。

1. 采用 RequestDispatcher 的方式进行

在 JSP 页面中添加如下代码：

```jsp
<%
response.setContentType("application/x-download");
    //设置为下载 application/x-download
String filedownload = "/要下载的文件名";//即将下载的文件的相对路径
String filedisplay = "最终要显示给用户的保存文件名";//下载文件时显示的文件保存名称
filenamedisplay = URLEncoder.encode(filedisplay,"UTF-8");
response.addHeader("Content-Disposition","attachment;filename="
                    + filedisplay);
try{
    RequestDispatcher dis = application.getRequestDispatcher
                        (filedownload);
    if(dis!= null) {
        dis.forward(request,response);
    }
    response.flushBuffer();
}catch(Exception e) {
    e.printStackTrace();
}finally{
}
%>
```

2. 采用文件流输出的方式下载

```jsp
<%@page language="java" contentType="application/x-msdownload"
    pageEncoding="utf-8"%><%
//关于文件下载时采用文件流输出的方式处理：
//加上 response.reset()，并且所有的%>后面不要换行，包括最后一个；
response.reset();//可以加也可以不加
response.setContentType("application/x-download");
String filedownload = "想办法找到要提供下载的文件的物理路径＋文件名";
String filedisplay = "给用户提供的下载文件名";
filedisplay = URLEncoder.encode(filedisplay,"UTF-8");
response.addHeader("Content-Disposition","attachment;filename="
                    + filedisplay);
OutputStream outp = null;
FileInputStream in = null;
try{
    outp = response.getOutputStream();
    in = new FileInputStream(filenamedownload);
    byte[] b = new byte[1024];
```

```
        int i = 0;
while((i = in.read(b)) > 0) {
    outp.write(b, 0, i);
}
outp.flush();
}
catch(Exception e) {
    System.out.println("Error!");
    e.printStackTrace();
}finally{
    if(in != null) {
        in.close();
        in = null;
    }
    if(outp != null) {
        outp.close();
        outp = null;
    }
}
%>
```

## 7.4 小　　结

本章主要介绍了文件属性信息的获取，文件和目录的删除，文件上传和下载操作。

## 7.5 习　　题

1. 设计一个 Web 应用，实现会员头像图片上传。
2. 学习使用开源框架 CommonFileUpload，完成文件上传。

# 第 8 章 JDBC 与数据库访问

本章主要介绍了数据库编程的定义，理解 JDBC 技术，熟悉 JDBC API，掌握使用 JDBC 访问数据库的方法、JSP 中常用数据库操作、JSP 中分页结束、JSP 中常见乱码问题及解决方案，讲解了应用实例等。

## 8.1 JDBC 基础知识

JDBC（Java DataBase Connectivity，Java 数据库连接）是面向应用程序开发人员和数据库驱动程序开发人员的应用程序接口（Application Programming Interface，API）。

### 8.1.1 JDBC 简介

#### 1. 什么是 JDBC

JDBC 是一个面向对象的应用程序接口，通过它可访问各类关系型数据库。JDBC 也是 Java 核心类库的一部分，由 Java 语言编写的类和界面组成。

自从 Java 语言于 1995 年 5 月正式公布以来，Java 语言风靡全球，出现大量用 Java 语言编写的程序，其中也包括数据库应用程序。由于没有一个 Java 语言的 API，编程人员不得不在 Java 程序中加入 C 语言的 ODBC（Open Database Connectivity）函数调用、这就使 Java 的很多优秀特性无法充分发挥，比如平台无关性、面向对象特性等。随着越来越多的编程人员对 Java 语言的喜爱，越来越多的公司在 Java 程序开发上投入的精力日益增加，对 Java 语言接口访问数据库的 API 要求越来越强烈。也由于 ODBC 的有其不足之处，比如它不容易使用，没有面向对象的特性等，Sun 公司决定开发以 Java 语言为接口的数据库应用程序开发接口。在 JDK1.x 版本中，JDBC 只是一个可选部件，到了 JDK1.1 公布时，SQL 类（也就是 JDBC API）就成为了 Java 语言的标准部件。

JDBC 给数据库应用开发人员、Java Web 开发人员提供了一种标准的应用程序设计接口，使开发人员可以用纯 Java 语言编写完整的数据库应用程序。通过使用 JDBC，开发人员可以很方便地将 SQL 语句传送给几乎任何一种数据库。也就是说，开发人员可以不必编写一个程序访问 MySQL，编写另一个程序访问 Oracle，再编写一个程序访问 Microsoft 的 SQL Server。用 JDBC 编写的程序能够自动地将 SQL 语句传送给相应的数据库管理系统（DBMS）。不但如此，使用 Java 编写的应用程序可以在任何支持 Java 的平台上运行，不必在不同的平台上编写不同的应用程序。Java 和 JDBC 的结合可以让开发人员在开发数据库应用时真正实现"Write Once，Run Everywhere！"

#### 2. JDBC 的作用

在数据库连接编程中，JDBC 的作用如下：

- 连接数据库；
- 发送 SQL 语句；
- 处理结果。

3. JDBC 驱动器实现方式

Java 数据库连接体系结构是用于 Java 应用程序连接数据库的标准方法。JDBC 对 Java 程序员而言是 API，对实现与数据库连接的服务提供商而言是接口模型。作为 API，JDBC 为程序开发提供标准的接口，并为数据库厂商及第三方中间件厂商实现与数据库的连接提供了标准方法。

在使用 JDBC 的 Java 应用程序中，客户端程序中只调用 JDBC 语句，真正与数据库通信是由数据库厂商提供的 JDBC 驱动器程序来完成的。

使用 JDBC 编程需要导入两个程序包：java.sql.*（核心 API）；javax.sql.*（扩展 API）。JDBC 驱动器根据其实现方式分为 4 种类型。

最初的 Java 语言规范并没有规定 Java 程序如何访问数据库。但不久之后，Sun 和它的合作者就开始填补这个空白。早期的 Java 数据访问策略依赖于建立通向 ODBC（ODBC 是 Microsoft 发起的数据源访问标准）的桥梁，结果就是 JDBC-ODBC 桥接驱动程序。今天，JDBC 驱动程序总共有 4 种类型。

（1）Sun 建议第一类驱动程序只用于原型开发，而不要用于正式的运行环境。桥接驱动程序由 Sun 提供，它的目标是支持传统的数据库系统。Sun 为该软件提供关键问题的补丁，但不为该软件的最终用户提供支持。一般地，桥接驱动程序用于已经在 ODBC 技术上投资的情形，例如已经投资了 Windows 应用服务器。

尽管 Sun 提供了 JDBC-ODBC 桥接驱动程序，但由于 ODBC 会在客户端装载二进制代码和数据库客户端代码，这种技术不适用于高事务性的环境。另外，这类 JDBC 驱动程序不支持完整的 Java 命令集，而是局限于 ODBC 驱动程序的功能。

（2）第 2 类 JDBC 驱动程序是本机 API 的部分 Java 代码的驱动程序，用于把 JDBC 调用转换成主流数据库 API 的本机调用。这类驱动程序也存在与第一类驱动程序一样的性能问题，即客户端载入二进制代码的问题，而且它们被绑定了特定的平台。

第 2 类驱动程序要求编写面向特定平台的代码，这对于任何 Java 开发者来说恐怕都不属于真正乐意做的事情。主流的数据库厂商，例如 Oracle 和 IBM，都为它们的企业数据库平台提供了第 2 类驱动程序，使用这些驱动程序的开发者必须及时跟进不同数据库厂商针对不同操作系统发行的各个驱动程序版本。

另外，由于第 2 类驱动程序没有使用纯 Java 的 API，把 Java 应用连接到数据源时，往往必须执行一些额外的配置工作。很多时候，第 2 类驱动程序不能在体系结构上与大型主机的数据源兼容，即使做到了兼容，效果也是差强人意。

由于诸如此类的原因，大多数 Java 数据库开发者选择第 3 类驱动程序，或者选择更灵活的第 4 类纯 Java 新式驱动程序。

（3）第 3 类 JDBC 驱动程序是面向数据库中间件的纯 Java 驱动程序，JDBC 调用被转换成一种中间件厂商的协议，中间件再把这些调用转换到数据库 API。第 3 类 JDBC 驱动程序的优点是它以服务器为基础，也就是不再需要客户端的本机代码，这使第 3 类驱动程序要比第 1、2 两类快。另外，开发者还可以利用单一的驱动程序连接到多种数据库。

（4）第 4 类 JDBC 驱动程序是直接面向数据库的纯 Java 驱动程序，即所谓的"瘦"（thin）驱动程序，它把 JDBC 调用转换成某种直接可被 DBMS 使用的网络协议，这样，客户机和应用服务器可以直接调用 DBMS 服务器。对于第 4 类驱动程序，不同 DBMS 的驱动程序不同。因此，在一个异构计算环境中，驱动程序的数量可能会比较多。但是，由于第 4 类驱动程序具有较高的性能，能够直接访问 DBMS，所以这一问题就不那么突出了。

第 3、4 两类都是纯 Java 的驱动程序，因此，对于 Java 开发者来说，它们在性能、可移植性、功能等方面都有优势。

4. JDBC API

JDBC 是一套数据库编程接口 API 函数，由 Java 语言编写的类、界面组成。JDBC 使用已有的 SQL 标准并支持与其他数据库连接标准，如 ODBC 之间的桥接。JDBC 实现了所有这些面向标准的目标，并且具有简单、严格类型定义且高性能实现的接口。用 JDBC 写的程序能够自动地将 SQL 语句传送给相应的数据库管理系统。

JDBC API 包括下列 5 部分：

（1）Driver 类。驱动程序，负责定位并访问数据库，建立数据库连接。

（2）DriverManager 类。负责处理驱动的调入并且对产生新的数据库连接提供支持。

（3）Connection 类。负责对特定数据库的连接。

（4）Statement 类。负责代表一个特定的容器，来对一个特定的数据库执行 SQL 语句。Statement 又有两个子类型：PreparedStatement 用于执行预编译的 SQL 语句，CallableStatement 用于执行对一个数据库内嵌过程的调用。

（5）ResultSet 类。负责控制对一个特定语句的行数据的存取。

5. JDBC 的结构

JDBC 结构如图 8-1 所示。

图 8-1  JDBC 结构

（1）用户应用程序实现 JDBC 的连接、发送 SQL、然后获取结果。它执行以下任务：请求与数据源建立连接；向数据源发送 SQL 请求；为结果集定义存储应用和数据类型；询问结果；过程处理错误；控制传输；提交操作；关闭连接。

（2）JDBC API 是一个标准统一的 SQL 数据存取接口。JDBC 的作用与 ODBC 作用类似，它为 Java 程序提供统一的操作各种数据库的接口，程序员编程时，可以不关心它所要操作的数据库是哪个厂家的产品，从而提高了软件的通用性。只要系统上安装了正确的驱动器组件，JDBC 应用程序就可以访问其相关的数据库。

（3）JDBC 驱动程序管理器的主要作用是代表用户的应用程序调入特定驱动程序，要完成的任务包括：为特定数据库定位驱动程序；处理 JDBC 初始化调用；为每个驱动程序提供 JDBC 功能入口点；为 JDBC 调用执行参数和结果有效性。

（4）驱动程序实现 JDBC 的连接，向特定数据源发送 SQL 声明，并且为应用程序获取结果。

（5）数据库由用户应用程序想访问的数据源和自身参数组成（即 DBMS 类型）。

## 8.1.2 DriverManager

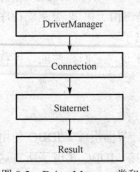

图 8-2 DriverManager 类和被管理对象间的关系

DriverManager 类负责管理数据库的连接，是 JDBC 的管理层，作用于用户和驱动程序之间。它跟踪可用的驱动程序，并在数据库和相应驱动程序之间建立连接。另外，DriverManager 类也处理诸如驱动程序登录时间限制及登录和跟踪消息的显示等事务。

DriverManager 类和被管理对象间的关系如图 8-2 所示。

DriverManager 类是 JDBC 的管理层，作用于用户和驱动程序间加载驱动程序，它所有的成员都是静态成员，所以在程序中无须对它进行实例化，直接通过类名就可以访问它。

DriverManager 类的方法如表 8-1 所示。

表 8-1  DriverManager 类的方法

| 方　法 | 说　明 |
|---|---|
| Connection getConnection(String url, String user,String password) | 建立对数据库的连接。url 的格式为：jdbc:subprotocol:subname，其中 jdbc 是保留字，subprotocol 指数据库类型，subname 指数据库位置；user 为连接数据库的用户名；password 为口令 |
| static Driver getDriver(String url) | 返回能够打开 url 所指定的数据库的驱动程序 |
| Class.forName(String driver) | 加载注册驱动程序 |

url 连接参数说明如表 8-2 所示。

表 8-2  连接参数说明

| 参 数 名 | 取　值 | 缺　省 |
|---|---|---|
| user | 数据库用户名 | 否 |
| password | 数据库用户口令 | 否 |
| autoReconnect | 当数据库连接丢失时是否自动连接，取值 true/false | false |
| maxReconnects | 如果 autoReconnect 为 true，则此参数为重试次数 | 3 |
| initialTimeout | 如果 autoReconnect 为 true，则此参数为重新连接前等待的秒数 | 2 |
| maxRows | 设置查询时返回的行数，0 表示全部 | 0 |
| useUnicode | 是否使用 unicode 输出，true/false | False |
| characterEncoding | 如果 useUnicode，则该参数制定 encoding 类型 | 否 |

**提示**：同时使用 useUnicode 和 characterEncoding，一般情况下可以解决使用数据库输出时的中文乱码问题，有关中文乱码的解决请参考 8.4.2。

对于简单的应用程序，一般程序员需要在此类中直接使用的唯一方法是 DriverManager.getConnection。正如名称所示，该方法将建立与数据库的连接。

使用 JDBC 驱动程序之前，必须先将驱动程序加载并向 DriverManager 注册后才可以使用，同时提供方法来建立与数据库的连接。

1. 加载 Driver 类

```
Class.forName("公司名.数据库名.驱动程序名，即驱动的名字")
```

例如：

```
Class.forName("com.mysql.jdbc.Driver")
```

2. 建立连接

加载 Driver 类并在 DriverManager 类注册后，就可用来与数据库建立连接。当调用 Driver.Manager.getConnection()发出连接请求时，DriverManager 将检查每个驱动程序，看它是否可以建立连接。

例如：以下是通常用驱动程序（JDBC 驱动程序）并连一个 student 数据源、用匿名登录的的示例：

```
Class.forName("com.mysql.jdbc.Driver");
```

3. 加载驱动程序

对于每种数据库，数据库厂商都提供有 JDBC 驱动，所以每种数据库的 JDBC 驱动都不一样，MySQL5 的驱动为 com.mysql.jdbc.Driver。

```
Connection conn=
DriverManager.getConnection("jdbc:mysql://localhost:3306/student",
    "root","root");
```

4. 建立连接

其中 URL 部分为 jdbc:mysql://localhost:3306/student，参数 localhost 是本机操作，如果数据库不在本机，可以改成数据所在机器上的 IP 地址；3306 是 MySQL 的端口地址，每种数据库都有自己的端口号；student 是数据库名；root 是 MySQL 的登录账号（用户名），第二个 root 是登录密码。

### 8.1.3 Connection

Connection 类负责维护 JSP/Java 数据库程序和数据库之间的连接。

作用是管理指向数据库的连接，如向数据库发送查询和接收数据库的查询结果都是在它的基础上的；完成同数据库的连接的所有任务之后关闭此连接。

Connection 类方法如表 8-3 所示。

表 8-3 Connection 类的方法

| 方　　法 | 说　　明 |
| --- | --- |
| Statement createStatement() | 建立 Statement 类对象 |
| Statement createStatement(int resultSetType,int resultSetConcurrency) | 建立 Statement 类对象，其中 resultSetType 值为：<br>TYPE_FORWARD_ONLY 结果集不可滚动<br>TYPE_SCROLL_INSENSITIVE 结果集可滚动，不反映数据库的变化<br>TYPE_SCROLL_SENSITIVE 结果集可滚动，反映数据库的变化 |
| boolean getAutoCommit() | 返回 Connection 类对象的 AutoCommit 状态 |
| void setAutoCommit(boolean autoCommit) | 设定 Connection 类对象的 AutoCommit 状态 |
| void commit() | 确定执行对数据库新增、删除或修改记录的操作 |
| void rollback() | 取消执行对数据库新增、删除或修改记录的操作 |
| void close() | 结束 Connection 对象对数据库的联机 |
| boolean isClosed() | 测试是否已经关闭 Connection 类对象对数据库的连接 |

### 8.1.4 Statement

Statement 类的作用是通过 Statement 类所提供的方法，可以利用标准的 SQL 命令对数据库直接新增、删除或修改操作。

1. 创建 Statement 对象

建立了到特定数据库的连接后，就可用该连接发送 SQL 语句。Statement 对象用 Connection 的方法 createStatement 创建。

例如：

```
Connection cn=DriverManager.getconnection(rul,"user","password");
Statement stmt=cn.createStatement();
```

Statement 对象 stmt 用于存放 SQL 语句使用。其中，creatStatement(int type,int concurrency) 方法有两个参数。type 参数有两个常量值：值为 ResultSet.TYPE_FORWARD_ONLY 时，当游标上下滚动，数据库变化时，当前结果集不变；值为 ResultSet.TYPE_SCROLL_INSENSITIVE 时，当游标上下滚动，数据库变化时，当前结果集随着变化。concurrency 参数有两个常量值：值为 ResultSet.CONCUR_READ_ONLY 时，不能用结果集更新数据库中的表；值为 ResultSet.CONCUR_UPDATABLE 时，可用结果集更新数据库中的表。

2. 执行 Statement 对象

为了执行 Statement 对象，被发送到数据库的 SQL 语句将被作为参数提供给 Statement 的方法。

例如：

```
ResultSet rs=stmt.executeQuery("select a,b,c from table1");
```

使用 Statement 对象执行语句，Statement 接口提供了 3 种执行 SQL 语句的方法：executeQuery()用于产生单个结果集的语句，如：select 语句；executeUpdate()用于执行 insert、update 或 delete 语句等，返回值是一个整数，指示受影响的行数（即更新计数）；execute()用于执行返回多个结果集、多个更新计数或二者组合的语句。

3. 语句完成

语句在已执行且所有结果返回时，即认为已完成。

对于返回一个结果集的 executeQuery()方法，在检索完 ResultSet 对象的所有行时该语句完成。对于方法 executeUpdate()，当它执行时语句即完成。在少数调用 execute()的情况下，只有在检索所有结果集或它生成的更新计数之后语句才完成。

4. 关闭 Statement 对象

Statement 对象将由 Java 垃圾收集程序自动关闭。但我们最好显式地关闭它们，因为这样会立即释放数据管理系统资源，有助于避免潜在的内存问题。

Statement 类的方法如表 8-4 所示。

表 8-4　Statement 类的方法

| 方　法 | 说　明 |
|---|---|
| ResultSet executeQuery(String sql) | 使用 SELECT 命令对数据库进行查询 |
| int executeUpdate(String sql) | 使用 INSERT\DELETE\UPDATE 对数据库进行新增、删除和修改操作 |
| void close() | 结束 Statement 类对象对数据库的连接 |

### 8.1.5　ResultSet

ResultSet 类负责存储查询数据库的结果，并提供一系列的方法对数据库进行新增、删除和修改操作。它也负责维护一个记录指针（Cursor），记录指针指向数据表中的某个记录，通过适当地移动记录指针，可以随心所欲地存取数据库，提高程序的效率。ResultSet 包含符合 SQL 语句中条件的所有行，且它通过一套 get 方法（这些 get 方法可以访问当前行中的不同列）提供了对这些行中数据的访问。

ResultSet 类的方法如表 8-5 所示。

表 8-5　ResultSet 类的方法

| 方　法 | 说　明 |
|---|---|
| boolean absolute(int row) | 移动记录指针到指定的记录 |
| void beforeFirst() | 移动记录指针到第一项记录之前 |
| void afterLast() | 移动记录指针到最后一项记录之后 |
| boolean first() | 移动记录指针到第一项记录 |
| boolean last() | 移动记录指针到最后一项记录 |
| boolean next() | 移动记录指针到下一项记录 |
| boolean previous() | 移动记录指针到上一项记录 |
| void deleteRow() | 删除记录指针指向的记录 |
| void moveToInsertRow() | 移动记录指针以新增一笔记录 |
| void moveToCurrentRow() | 移动记录指针到被记忆的记录 |
| void insertRow() | 新增一项记录到数据库中 |
| void updateRow() | 修改数据库中的一项记录 |
| void update 类型(int columnIndex,类型 x) | 修改指定字段的值 |
| int get 类型(int columnIndex) | 取得指定字段的值 |
| ResultSetMetaData getMetaData() | 取得 ResultSetMetaData 类对象 |

注：记录结果集是一张二维表，其中有查询所返回的列标题及相应的值。

## 8.2　使用 JDBC 访问数据库

使用 JDBC-ODBC 桥连接可以连接 Access、MySQL、Microsoft SQL Server 和 Oracle 等数据库。建议尽可能使用纯 Java JDBC 驱动程序代替桥和 ODBC 驱动程序，这可以完全省去 ODBC 所需的客户机配置，也免除了 Java 虚拟机被桥引入的本地代码中的错误所破坏的可能性。

其实每个数据库厂商都提供了数据库的 JDBC 驱动程序，可以使用 DBMS 厂商提供的 JDBC 驱动访问数据库。下面就分别介绍通过 JDBC 驱动访问 MySQL、Microsoft SQL Server 数据库。

### 8.2.1　使用 JDBC 访问数据库的一般步骤

利用 JDBC API 进行 Java 应用的数据库编程的基本流程：取得数据库连接；执行 SQL 语句；处理执行结果；释放数据库连接。

1. 取得数据库连接

用 DriverManager 取数据库连接。

例子：

```
String className, url, uid, pwd;
className ="com.microsoft.jdbc.sqlserver.SQLServerDriver";
url = "jdbc:microsoft:sqlserver://localhost:1433;DatabaseName=pubs";
uid = "sa";
pwd = "admin";
Class.forName(className);
Connection con = DriverManager.getConnection(url,uid,pwd);
```

2.执行 sql 语句

（1）用 Statement 来执行 SQL 语句。

```
String sqlStr;
Statement stm = con.createStatement();
stm.executeQuery(sql);   // 执行数据查询语句（select）
stm.executeUpdate(sql);
// 执行数据更新语句（delete、update、insert、drop 等）
stm.close();
```

（2）用 PreparedStatement 来执行 SQL 语句。

```
String sql;
sql = "insert into user (id,name) values (?,?)";
PreparedStatement ps = cn.prepareStatement(sql);
ps.setInt(1,xxx);
ps.setString(2,xxx);
...
ResultSet rs = ps.executeQuery();    // 查询
int i = ps.executeUpdate();          // 更新
```

3. 处理执行结果

查询语句，返回记录集 ResultSet。

更新语句，返回数字，表示该更新影响的记录数。

ResultSet 的方法如下：

（1）next()，将记录指针往后移动一行，如果成功返回 true；否则返回 false。

（2）getInt("id")或 getSting("name")，返回当前记录指针下某个字段的值。

（3）释放连接。

4. 释放数据库连接

一般先关闭 ResultSet，然后关闭 Statement（或者 PreparedStatement）；最后关闭 Connection。

```
rs.close();
stm.close();
con.close();
```

5. 总结

使用 JDBC API 操作数据库的具体编程步骤如下：

（1）建立数据库连接，获得 Connection 对象。
（2）根据用户的输入组装 SQL 查询语句。
（3）根据 SQL 语句建立 Statement 对象或者 PreparedStatement 对象。
（4）用 Connection 对象执行 SQL 语句，获得结果集 ResultSet 对象。
（5）一条一条读取结果集 ResultSet 对象中的数据。
（6）根据读取到的数据，按特定的业务逻辑进行计算。
（7）依次关闭各个 Statement 对象和 Connection 对象。

### 8.2.2　使用 JDBC 驱动访问 MySQL 数据库

**1. MySQL JDBC 驱动下载和配置**

本节使用的是 MySQL 5.5，下载支持 5.5 版本的 JDBC 驱动。使用的 JDBC 驱动是 mysql-connector-java-5.1.18-bin.jar。下载完成 MySQL 的 JDBC 驱动后，把该文件可以放到任意目录下，这里假设该目录是"D:\JSP 程序设计"。设置 ClassPath 以保证 Web 服务器能够访问到这个驱动程序。设置方法如下：

（1）依次进入"我的电脑"->"系统"->"设置环境变量"，如果已经有 ClassPath 变量就编辑该变量的值，否则增加 ClassPath 变量。
（2）设置 ClassPath 的值如下：

```
.;D:\JSP 程序设计\mysql-connector-java-5.1.18-bin.jar
```

（3）重新启动服务器即可。

如果开发 JSP 项目使用的是 Eclipse 和 NetBeans 开发工具，配置如下：

（1）MySQL 的 JDBC 驱动在 Eclipse 中的配置。首先把 mysql-connector-java-5.1.18-bin.jar 复制到 WEB-INF 中的 lib 文件夹中，另外在 Eclipse 项目 ch09 上右单击，在属性菜单中单击 Build Path->Configure Build Path，如图 8-3 所示。

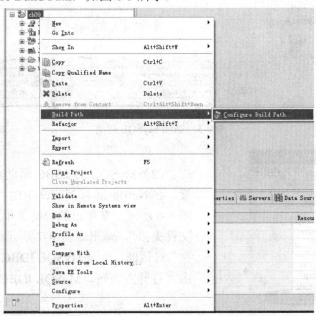

图 8-3　选择 Configure Build Path

弹出对话框如图 8-4 所示。在对话框中选择卡 Libraries->Add External JARs。

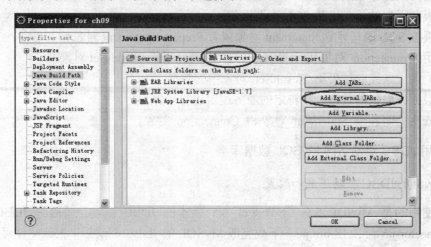

图 8-4 选择 Add External JARs

弹出对话框如图 8-5 所示。找到保存在项目 WEB-INF 中的 lib 文件夹 MySQL JDBC 驱动所在位置，找到驱动位置后单击"打开"按钮，MySQL JDBC 驱动在 Eclipse 中配置完成。

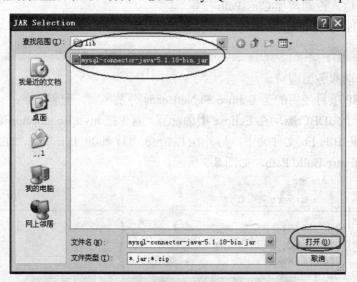

图 8-5 添加 MySQL 驱动

图 8-6 选择"添加 JAR/文件夹…"

（2）MySQL 的 JDBC 驱动在 NetBeans 中的配置。在 NetBeans 项目 ch9 中的"库"上右单击，如图 8-6 所示。在弹出的快捷菜单上单击"添加 JAR/文件夹…"，弹出如图 8-7 所示的"添加 JAR/文件夹"对话框。找到 MySQL JDBC 驱动所在位置后单击"打开"按钮，MySQL JDBC 驱动配置完成。

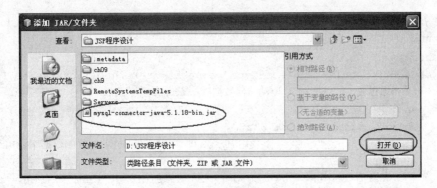

图 8-7 添加 MySQL 驱动

## 2. MySQL 建立数据库和表

安装完 MySQL 以后，安装了一个 MySQL 的插件"Navicat V8.2.12ForMySQL 简体中文绿色特别版.exe"，该插件能够在使用 MySQL 时提供可视化、友好的图形用户界面。

使用 MySQL 建立数据库 student 和表 sinfo。数据库、表以及表的字段和类型如图 8-8 所示。

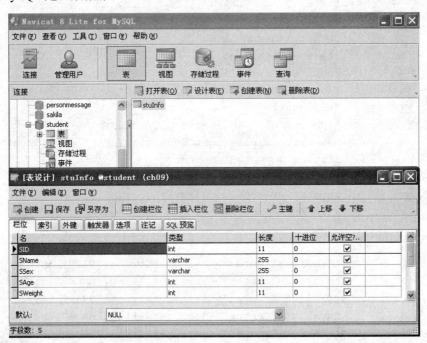

图 8-8 数据库、表以及字段

## 3. 编写 JSP 文件访问数据库（accessMySQL.jsp）

**例 8-1**：使用 JDBC 驱动访问 MySQL 的 JSP 页面。

accessMySQL.jsp

```
<%@page import="java.sql.*"%>
<%@ page language="java" contentType="text/html; charset=UTF-8"
    pageEncoding="UTF-8"%>
<!DOCTYPE html PUBLIC "-//W3C//DTD HTML 4.01 Transitional//EN"
```

```html
"http://www.w3.org/TR/html4/loose.dtd">
<html>
<head>
<meta http-equiv="Content-Type" content="text/html; charset=UTF-8">
<title>通过 MySQL 的 JDBC 驱动访问数据库</title>
</head>
    <body bgcolor="CCEEFF">
        <h2 align="center">使用 MySQL 的 JDBC 驱动访问 MySQL 数据库</h2>
        <hr>
        <table width="50%" border="2" bgcolor="#CCCEEE" align="center">
           <tr>
                <th align="center">学号</th>
                <th align="center">姓名</th>
                <th align="center">性别</th>
                <th align="center">年龄</th>
                <th align="center">体重（公斤）</th>

           </tr>
           <%
                Connection con=null;
                Statement stmt=null;
                ResultSet rs=null;
                Class.forName("com.mysql.jdbc.Driver");
                String url=
                "jdbc:mysql://localhost:3306/student
                con=DriverManager.getConnection(url,"root","root");
                stmt=con.createStatement();
                String sql="select * from stuinfo";
                rs=stmt.executeQuery(sql);
                while(rs.next()){
          %>
          <tr>
                <td align="center"><%=rs.getString("SID")%></td>
                <td align="center"><%=rs.getString("SName")%></td>
                <td align="center"><%=rs.getString("SSex")%></td>
                <td align="center"><%=rs.getString("SAge")%></td>
                <td align="center"><%=rs.getString("SWeight")%></td>
          </tr>
          <%
                }
                rs.close();
                stmt.close();
                con.close();
          %>
        </table>
        <hr>
    </body>
</html>
```

accessMySQL.jsp 运行效果如图 8-9 所示。

图 8-9　accessMySQL.jsp 运行效果

### 8.2.3　访问 Microsoft SQL Server 2000 数据库及其应用实例

1. Microsoft SQL Server JDBC 驱动下载和配置

数据库使用的是 Microsoft SQL Server 2000，在 Microsoft 网站下载 SQL Server 2000 Driver for JDBC。下载后进行安装，在安装路径下有 3 个 JAR 包，如图 8-10 所示。

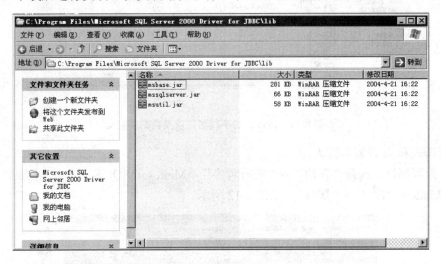

图 8-10　SQL Server 驱动所需的 JAR 文件

设置 ClassPath 以保证 Web 服务器能够访问到这个驱动程序。设置方法如下：
（1）依次进入"我的电脑"->"系统"->"设置环境变量"，如果已经有 ClassPath 变量就编辑该变量的值，否则增加 ClassPath 变量。
（2）设置 ClassPath 的值如下。

.;C:\Program Files\Microsoft SQL Server 2000 Driver for JDBC\lib\ msbase.jar;C:\Program Files\Microsoft SQL Server 2000 Driver for JDBC\lib\ mssqlserver.jar; C:\Program Files\Microsoft SQL Server 2000 Driver for JDBC\lib\ msutil.jar

（3）重新启动服务器即可。

如果开发 JSP 项目使用的是 Eclipse 或 NetBeans，加载 JDBC 驱动和加载 MySQL 的 JDBC 驱动相似。

在使用 Microsoft SQL Server 2000 JDBC 驱动时，需要加载 lib 包下的 3 个 JAR 文件，这里不再介绍。

2. Microsoft SQL Server 建立数据库和表

本例使用 SQL Server 2000 自带的 pubs 数据库。把登录模式设置为 Windows 和 SQL 混合登录模式，用户名为 sa，密码为空。查询 pubs 中的 jobs 表，表中的字段是：job_id、job_desc、min_lvl、max_lvl。数据库、表以及表的字段和数据如图 8-11 所示。

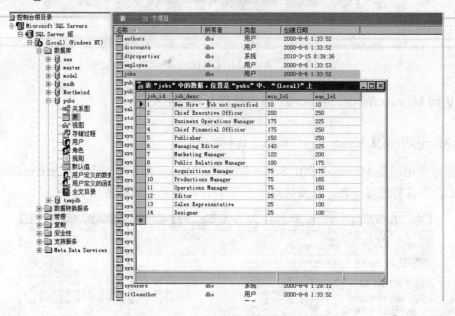

图 8-11　jobs 表的字段名称以及数据类型

登录模式和 sa 设置步骤为：

（1）配置属性。选择"开始"->"所有程序"->Microsoft SQL Server->"企业管理器"->"(local)(Windows NT)"->"属性"，如图 8-12 所示。

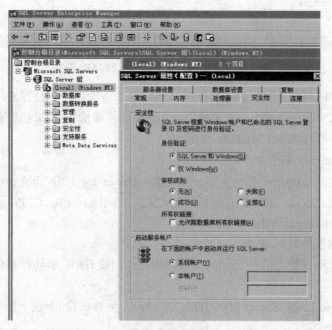

图 8-12　属性配置

（2）设置混合登录模式。在图 8-12 中，在"身份验证"中选择 SQL Server 和 Window(s)。

（3）添加系统管理员 sa。sa 是 SQL Server 默认的数据库管理员用户名。在图 8-12 左侧，选择"安全性"->"登录"->"新建登录"，如图 8-13 所示。单击"新建登录"后弹出如图 8-14 所示对话框，在该对话框中输入用户名为 sa，身份验证选择"SQL Server 身份验证"，为用户输入密码，选择要操作的数据库名 pubs。

（4）设置其他选项卡。在图 8-14 服务器角色选项卡中，选择数据库管理员身份为 System Administrators，如图 8-15 所示。

在图 8-15 "数据库访问"选项卡中，选择要访问的数据库名，并指定对该数据库所允许的操作，public 是公共操作，db_owner 是数据库所有者能够进行的操作，如图 8-16 所示。

图 8-13　添加登录

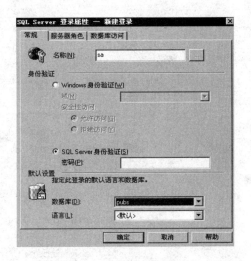

图 8-14　输入登录信息

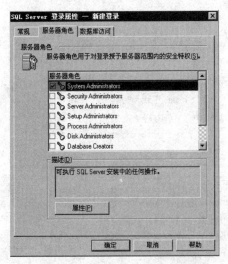

图 8-15　服务器角色选项卡

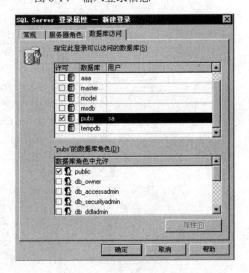

图 8-16　数据库访问选项卡

3. 编写 JSP 文件访问数据库

**例 8-2**：使用驱动访问 SQL Server 2000 的 JSP 页面。

accessSQLServer.jsp

```jsp
<%@page import="java.sql.*"%>
<%@ page language="java" contentType="text/html; charset=UTF-8"
    pageEncoding="UTF-8"%>
<!DOCTYPE html PUBLIC "-//W3C//DTD HTML 4.01 Transitional//EN"
 "http://www.w3.org/TR/html4/loose.dtd">
<html>
<head>
<meta http-equiv="Content-Type" content="text/html; charset=UTF-8">
 <title>通过 SQL Server 的 JDBC 驱动访问数据库</title>
</head>
    <body>
        <center>
          <br> <br> <br>
          <h2> 欢迎使用 SQL Server 的 JDBC 驱动访问 SQL Server 数据库</h2>
          <hr>
          <table border=2 bgcolor="ccceee" align="center">
            <tr>
                <td>job_id</td>
                <td>job_desc</td>
                <td>min_lvl</td>
                <td>max_lvl</td>
            </tr>
            <%
            Class.forName("com.microsoft.jdbc.sqlserver.SQLServerDriver");
                String
              url="jdbc:microsoft:sqlserver://localhost:1433;
                DatabaseName=pubs";
                String user="sa";
                String password="";
                Connection conn= DriverManager.getConnection(url,
                  user,password);
                Statement stmt=conn.createStatement();
                String sql="select top 6 * from jobs";
                ResultSet rs=stmt.executeQuery(sql);
                while(rs.next()){
            %>
              <tr>
                <td><%=rs.getString("job_id")%> </td>
                <td><%=rs.getString("job_desc")%> </td>
                <td><%=rs.getString("min_lvl")%> </td>
                <td><%=rs.getString("max_lvl")%> </td>
              </tr>
            <%
```

```
                }
                rs.close();
                stmt.close();
                conn.close();
            %>
        </table>
        <hr>
      </center>
   </body>
</html>
```

accessSQLServer.jsp 运行效果如图 8-17 所示。

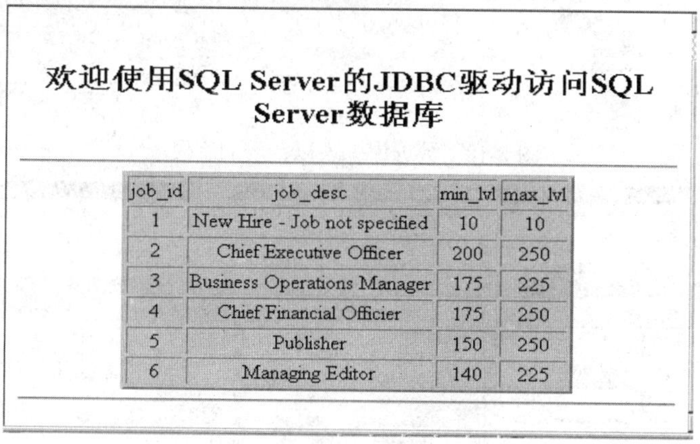


| job_id | job_desc | min_lvl | max_lvl |
|---|---|---|---|
| 1 | New Hire - Job not specified | 10 | 10 |
| 2 | Chief Executive Officer | 200 | 250 |
| 3 | Business Operations Manager | 175 | 225 |
| 4 | Chief Financial Officier | 175 | 250 |
| 5 | Publisher | 150 | 250 |
| 6 | Managing Editor | 140 | 225 |

图 8-17　accessSQLServer.jsp 运行效果

### 8.2.4 访问 Microsoft SQL Server 2008 数据库及其应用实例

#### 1. Microsoft SQL Server JDBC 驱动下载和配置

如果数据库使用的是 Microsoft SQL Server 2008，在 Microsoft 网站下载 Microsoft JDBC Driver 4.0 for SQL Server。下载后进行解压，解压后有 2 个 JAR 包。设置 ClassPath 以保证 Web 服务器能够访问到这个驱动程序。设置方法如下：

（1）依次进入“我的电脑”->“系统”->“设置环境变量”，如果已经有 ClassPath 变量就编辑该变量的值，否则增加 ClassPath 变量。

（2）设置 ClassPath 的值如下。

.;D:\JSP 程序设计\ch09\Microsoft JDBC Driver 4.0 for SQL Server\sqljdbc_4.0\chs\sqljdbc4.jar

（3）重新启动服务器即可。

如果开发 Java Web 项目使用的是 Eclipse 或 NetBeans，加载 JDBC 驱动和加载 MySQL 的驱动相似，不过在使用 Microsoft JDBC Driver 4.0 for SQL Server 驱动时，需要 sqljdbc4.jar 文件。

#### 2. Microsoft SQL Server 2008 建立数据库和表

本例中使用 SQL Server 2008 建一个 student 数据库，并在该数据库中创建一个 info 表，数据库、表以及表的字段如图 8-18 所示。表的数据如图 8-19 所示。

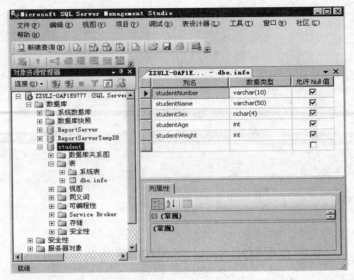

图 8-18 数据库、表以及表的字段

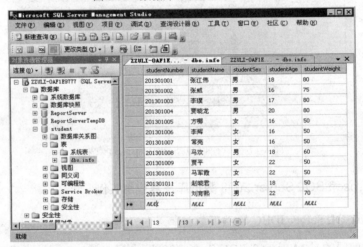

图 8-19 表中数据

登录模式和 sa 设置步骤如下：

（1）选择"开始"->"所有程序"->Microsoft SQL Server 2008->SQL Server Management Studio，如图 8-20 所示。弹出如图 8-21 所示的"连接到服务器"对话框。选择服务器名称和身份验证后单击"连接"按钮，弹出如图 8-22 所示的 SQL Server Management Studio 管理界面。右击图 8-22 中服务器的属性，如图 8-23 所示。弹出如图 8-24 所示的"服务器属性"对话框。

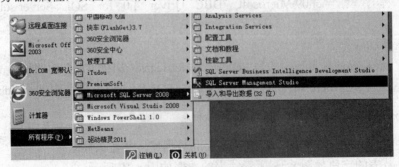

图 8-20 启动 SQL Server 2008 管理

图 8-21 连接服务器

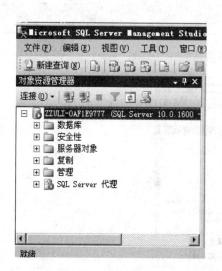

图 8-22 管理界面

图 8-23 选择服务器属性

（2）设置混合登录模式。在图 8-24 中，单击"选择页"中的"安全性"后，选择"SQL Server 和 Windows 身份验证模式"。

（3）登录设置。sa 是 SQL Server 默认的数据库管理员用户名。在图 8-22 中服务器属性结构中，选择"安全性"->"登录"->"属性"，如图 8-25 所示，弹出如图 8-26 所示对话框，输入密码，选择要操作的数据库名 student。另外，将 8-26 中的"状态"选项改为启用。

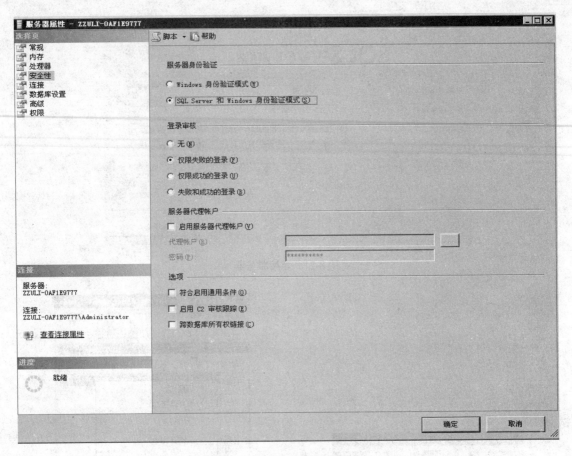

图 8-24 服务器属性对话框

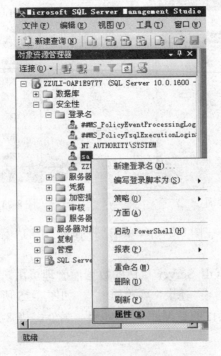

图 8-25 配置属性

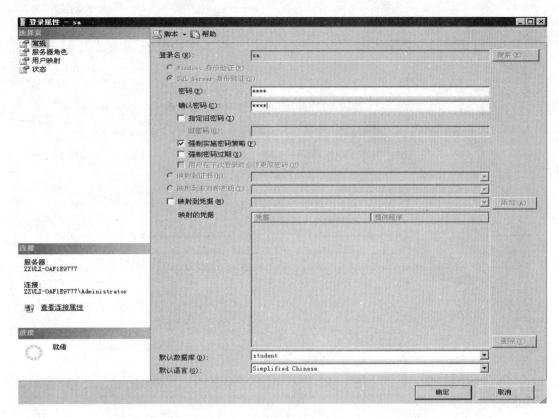

图 8-26 设置密码并指定数据库

3. 编写 JSP 文件访问数据库（accessSQLServer2008.jsp）

例 8-3：使用驱动访问 SQL Server 2008 的 JSP 页面。

accessSQLServer2008.jsp

```
<%@page import="java.sql.*"%>
<%@ page language="java" contentType="text/html; charset=UTF-8"
    pageEncoding="UTF-8"%>
<!DOCTYPE html PUBLIC "-//W3C//DTD HTML 4.01 Transitional//EN"
 "http://www.w3.org/TR/html4/loose.dtd">
<html>
<head>
<meta http-equiv="Content-Type" content="text/html; charset=UTF-8">
 <title>通过 JDBC 驱动访问 SQL Server 2008</title>
</head>
    <body>
        <center>
         <br> <br> <br>
        <h2>使用 JDBC 驱动访问 SQL Server 2008 数据库</h2>
        <hr>
        <table border=2 bgcolor="ccceee" align="center">
            <tr>
                <th>学号</th>
                <th>姓名</th>
                <th>性别</th>
```

```
            <th>年龄</th>
            <th>体重</th>

        </tr>
        <%
            //2008 的 JDBC 驱动和 URL 与 2000 的驱动和 URL 不一样，请比较
            Class.forName("com.microsoft.sqlserver.jdbc.SQLServerDriver");
           String
           url="jdbc:sqlserver://localhost:1433;databasename=student";
            String  user="sa";
            String  password="root";
            Connection  conn=  DriverManager.getConnection(url,
              user,password);
            Statement  stmt=conn.createStatement();
            String  sql="select * from info";
            ResultSet  rs=stmt.executeQuery(sql);
            while(rs.next()){
        %>
        <tr>
            <td><%=rs.getString("studentNumber")%> </td>
            <td><%=rs.getString("studentName")%> </td>
            <td><%=rs.getString("studentSex")%> </td>
            <td><%=rs.getString("studentAge")%> </td>
            <td><%=rs.getString("studentWeight")%> </td>
        </tr>
        <%}%>
        <%
             rs.close();
             stmt.close();
             conn.close();
        %>
      </table>
       <hr>
     </center>
   </body>
</html>
```

accessSQLServer2008.jsp 运行效果如图 8-27 所示。

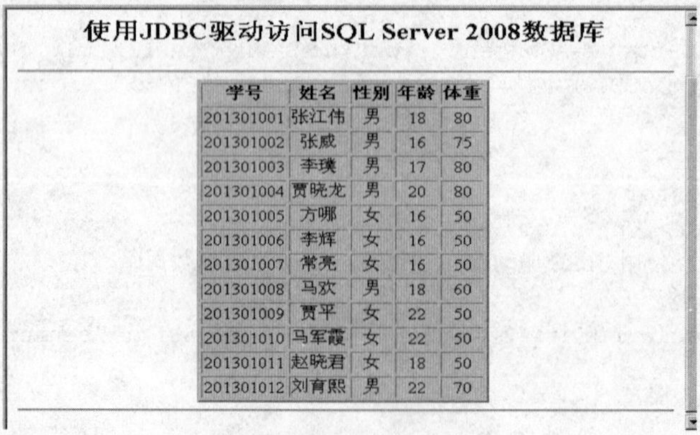


| 学号 | 姓名 | 性别 | 年龄 | 体重 |
|---|---|---|---|---|
| 201301001 | 张江伟 | 男 | 18 | 80 |
| 201301002 | 张威 | 男 | 16 | 75 |
| 201301003 | 李璞 | 男 | 17 | 80 |
| 201301004 | 贾晓龙 | 男 | 20 | 80 |
| 201301005 | 方哪 | 女 | 16 | 50 |
| 201301006 | 李辉 | 女 | 16 | 50 |
| 201301007 | 常亮 | 女 | 16 | 50 |
| 201301008 | 马欢 | 男 | 18 | 60 |
| 201301009 | 贾平 | 女 | 22 | 50 |
| 201301010 | 马军霞 | 女 | 22 | 50 |
| 201301011 | 赵晓君 | 女 | 18 | 50 |
| 201301012 | 刘育熙 | 男 | 22 | 70 |

图 8-27　accessSQLServer2008.jsp 运行效果

## 8.3 数据库的增、删、改、查操作

### 8.3.1 数据库的增、删、改、查操作

#### 1. 添加操作

在 SQL 语句中，使用 INSERT 语句将新行添加到表或视图中，语法格式如下：

```
INSERT INTO table_name column_list VALUES({DEFAULT|NULL|expression} [,…n]);
```

其中，table_name 指定将要插入数据的表或 table 变量的名称；column_list 是要在其中插入数据的一列或多列的列表。必须用圆括号将 column_list 括起来，并且用逗号进行分隔；VALUES ( { DEFAULT | NULL | expression } [,…n] )为要插入数据值的列表。对 column_list（如果已指定）中或者表中的每个列，都必须有一个数据值，且必须用圆括号将值列表括起来。如果 VALUES 列表中的值与表中列的顺序不相同，或者未包含表中所有列的值，那么必须使用 column_list 明确地指定存储每个传入值的列。

例如，在学生信息表中添加一个学生的信息('00001','david','male')，则对应的 SQL 语句应如下

```
INSERT INTO stuinfo values('00001','david','male');
```

#### 2. 删除操作

在 SQL 语言中，使用 DELETE 语句删除数据表中的行。DELETE 语句的语法格式如下：

```
DELETE FROM table_name [WHERE search_condition];
```

其中，table_name 用来指定表；WHERE 指定用于限制删除行数的条件。如果没有提供 WHERE 子句，则删除表中的所有记录。

例如，要从学生信息表中删除学号为"00001"的学生信息，则对应的 SQL 语句如下：

```
DELETE FROM stuInfo WHERE 学号='000001';
```

#### 3. 修改操作

SQL 语言中的更新语句是 UPDATE 语句，其语法格式如下：

```
UPDATE table_name SET column_name=expression[,column_name1=expression]
[WHERE search_condition];
```

其中，table_name 用来指定需要更新表的名称；SET column_name=expression[,column_name1=expression]指定要更新的列或变量名称的列表，column_name 指定含有要更改数据列的名称；WHERE search_condition 指定条件来限定所要更新的行。

例如，修改所有学生的年龄，将年龄都增加 1 岁，则对应的 SQL 语句如下：

```
UPDATE stuinfo SET 年龄=年龄+1;
```

#### 4. 查询操作

数据查询是数据库的一项基本操作，主要利用 SQL（Structure Query language，结构化查

询语言)语句和 ResultSet 对记录进行查询。查询数据库的方法有很多,可以分为顺序查询、随机查询、带参数查询等。

SQL 语句是标准的结构化查询语言,可以在任何数据库管理系统中使用,因此被普遍使用,其语法格式如下:

```
SELECT list FROM table
[WHERE search_condition]
[GROUP BY group_by_expression] [HAVING search_condition]
[ORDER BY order_expression [ASC/DESC] ];
```

各参数的含义如下。

- list:目标列表达式。用来指明要查询的列名,或是有列名参与的表达式。用*代表所有列。
- table:指定要查询表的名称。它可以是一张表也可以是多张表。如果不同表中有相同列,需要利用"表名.列名"的方式指明该列来自哪张表。
- search condition:查询条件表达式,用来设定查询的条件。
- group_by _expression:分组查询表达式。按表达式条件将记录分为不同的记录组参与运算,通常与目标列表达式中的函数配合使用,实现分组统计的功能。
- order_expression:排序查询表达式。按指定表达式的值来对满足条件的记录进行排序,默认是升序(ASC)。

SQL 中的查询语句除了可以实现单表查询以外,还可以实现多表查询和嵌套查询,使用起来比较灵活,完成较复杂的查询请参看其他资料。

JDBC 提供 3 种接口实现 SQL 语句的发送执行,分别是 Statement、PreparedStatement 和 CallableStatement。Statement 接口的对象用于执行简单的、不带参数的 SQL 语句;PreparedStatement 接口的对象用于执行带有 IN 类型参数的、预编译过的 SQL 语句;CallableStatement 接口的对象用于执行一个数据库的存储过程。PreparedStatement 继承了 Statement,而 CallableStatement 又从 PreparedStatement 继承而来。通过上述对象执行发送 SQL 语句,结果集由 JDBC 提供的 ResultSet 接口对结果集中的数据进行操作。下面分别对 JDBC 中执行发送 SQL 语句以及对执行过 SQL 语句的结果集操作的接口进行介绍。

使用 Statement 类发送要执行 SQL 语句前首先要创建 Statement 对象实例,然后根据参数 type、concurrency 的取值情况返回 Statement 类型的结果集。语法格式如下:

```
Statement stmt = con.createStatement(type,concurrency);
```

其中,type 属性用来设置结果集的类型。type 属性有 3 种取值:取值为 ResultSet.TYPE_FORWORD_ONLY 时,代表结果集的记录指针只能向下滚动;取值为 ResultSet.TYPE_SCROLL_INSENSITIVE 时,代表结果集的记录指针可以上下滚动,数据库变化时,当前结果集不变;取值为 ResultSet.TYPE_SCROLL_SENSITIVE 时,代表结果集的记录指针可以上下滚动,数据库变化时,结果集随之变动。

concurrency 属性用来设置结果集更新数据库的方式。它也有两种取值:当 concurrency 属性取值为 ResultSet.CONCUR_READ_ONLY 时,代表不能用结果集更新数据库中的表;而当 concurrency 属性的取值为 ResultSet.CONCUR_UPDATETABLE 时,代表可以更新数据库。

PreparedStatement 类可以将 SQL 语句传给数据库做预编译处理,即在执行的 SQL 语句中包含一个或多个 IN 参数，可以通过设置 IN 参数值多次执行 SQL 语句，不必重新给出 SQL 语句，这样可以大大提高执行 SQL 语句的速度。

所谓 IN 参数就是指那些在 SQL 语句创立时尚未指定值的参数，在 SQL 语句中 IN 参数用"?"号代替。

例如：

```
PreparedStatement pstmt=connection.preparedStatement("SELECT*FROM student WHERE 年龄>=? AND 性别=? ");
```

这个 PreparedStatement 对象用来查询表中指定条件的信息，在执行查询之前必须对每个 IN 参数进行设置，设置 IN 参数的语法格式如下：

```
pstmt.setXXX(position,value);
```

其中，XXX 为设置数据的各种类型，position 为 IN 参数在 SQL 语句中的位置，value 指该参数被设置的值。

例如：

```
pstmt.setInt(1,20);
```

### 8.3.2 基于 MVC 模式的学生信息管理系统

本节介绍一个综合应用实例，说明如何对数据库进行操作。本实例用到的数据库及其表与 8.2.2 节图 8-8 一样。

**1. 学生信息管理系统主页面功能的实现**

本实例有一个学生信息管理系统主页面（stuAdmin.jsp），该页面提供对学生信息的基本操作。主页面提供学生信息查询、学生信息添加、学生信息删除和学生信息修改功能，页面都存放在 studentManage 文件夹中。主页面如图 8-28 所示。

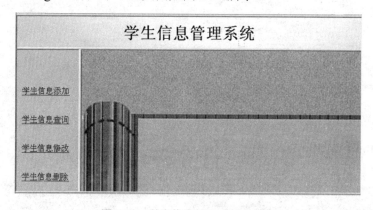

图 8-28 学生信息管理系统主页面

**例 8-4**：系统主页面。

stuAdmin.jsp

```
<%@ page language="java" contentType="text/html; charset=UTF-8"
    pageEncoding="UTF-8"%>
<!DOCTYPE html PUBLIC "-//W3C//DTD HTML 4.01 Transitional//EN"
```

```
    "http://www.w3.org/TR/html4/loose.dtd">
<html>
<head>
<meta http-equiv="Content-Type" content="text/html; charset=UTF-8">
<title>学生信息管理系统</title>
</head>
    <frameset rows="90,*">
        <frame src="../studentManage/top.jsp" scrolling="no">
        <frameset cols="126,*">
            <frame src="../studentManage/left.jsp" scrolling="no">
            <frame src="../studentManage/bottom.jsp" name="main"
            scrolling="no">
        </frameset>
    </frameset>
</html>
```

主页面 stuAdmin.jsp 使用框架由 3 个 JSP 页面构成，3 个 JSP 页面分别是：top.jsp、left.jsp 和 bottom.jsp。3 个 JSP 页面代码分别如下。

例 8-5：top.jsp 页面。

top.jsp

```
<%@ page language="java" contentType="text/html; charset=UTF-8"
    pageEncoding="UTF-8"%>
<!DOCTYPE html PUBLIC "-//W3C//DTD HTML 4.01 Transitional//EN"
 "http://www.w3.org/TR/html4/loose.dtd">
<html>
<head>
<meta http-equiv="Content-Type" content="text/html; charset=UTF-8">
<title>top</title>
</head>
    <body background="../image/top.jpg" >
        <center>
            <h1>学生信息管理系统</h1>
        </center>
    </body>
</html>
```

例 8-6：left.jsp 页面。

left.jsp

```
<%@ page language="java" contentType="text/html; charset=UTF-8"
    pageEncoding="UTF-8"%>
<!DOCTYPE html PUBLIC "-//W3C//DTD HTML 4.01 Transitional//EN"
 "http://www.w3.org/TR/html4/loose.dtd">
<html>
<head>
<meta http-equiv="Content-Type" content="text/html; charset=UTF-8">
<title>left</title>
</head>
<body bgcolor="CCCFFF">
```

```html
            <br><br><br>
            <p>
                <a href="addStudent.jsp" target="main">学生信息添加</a>
            </p>
            <br>
            <p>
                <a href="../LookStudentServlet" target="main">学生信息查询</a>
            </p>
            <br>
            <p>
                <a href="lookUpdateStudent.jsp" target="main">学生信息修改</a>
            </p>
            <br>
            <p>
                <a href="../LookDeleteStudentServlet" target="main">学生信息删除</a>
            </p>
    </body>
</html>
```

**例 8-7**：bottom.jsp 页面。

bottom.jsp

```jsp
<%@ page language="java" contentType="text/html; charset=UTF-8"
    pageEncoding="UTF-8"%>
<!DOCTYPE html PUBLIC "-//W3C//DTD HTML 4.01 Transitional//EN"
 "http://www.w3.org/TR/html4/loose.dtd">
<html>
<head>
<meta http-equiv="Content-Type" content="text/html; charset=UTF-8">
<title> bottom </title>
</head>
    <body background="../image/bottom.jpg">
    </body>
</html>
```

2. 学生信息添加功能的实现

单击图 8-28 中的"学生信息添加"链接出现如图 8-29 所示的页面（参考 left.jsp 中的"<a href="addStudent.jsp" target="main">学生信息添加</a>"）。超链接页面是 addStudent.jsp。

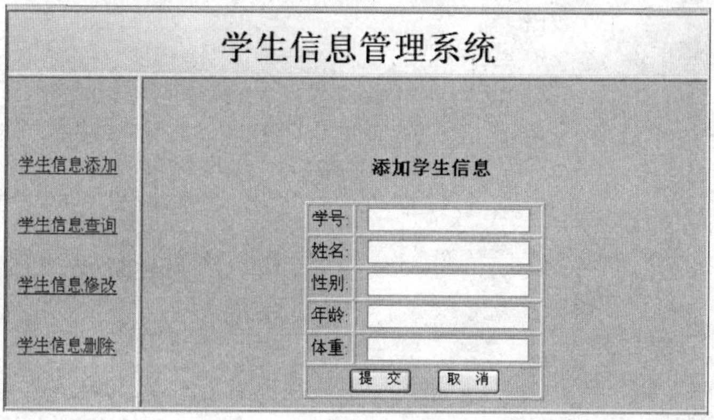

图 8-29　学生信息添加

例 8-8: addStudent.jsp 页面。

addStudent.jsp

```jsp
<%@ page language="java" contentType="text/html; charset=UTF-8"
    pageEncoding="UTF-8"%>
<!DOCTYPE html PUBLIC "-//W3C//DTD HTML 4.01 Transitional//EN"
 "http://www.w3.org/TR/html4/loose.dtd">
<html>
<head>
<meta http-equiv="Content-Type" content="text/html; charset=UTF-8">
<title>学生信息添加</title>
</head>
    <body bgcolor="CCCFFF">
        <center>
            <br><br><br>
            <h3>  添加学生信息</h3>
            <form action="../AddStudentServlet"  method="get">
                <table border="1" width="230">
                    <tr>
                        <td>学号:</td>
                        <td><input type="text" name="studentNumber"/></td>
                    </tr>
                    <tr>
                        <td>姓名:</td>
                        <td><input type="text" name="studentName"/></td>
                    </tr>
                    <tr>
                        <td>性别:</td>
                        <td><input type="text" name="studentSex"/></td>
                    </tr>
                    <tr>
                        <td>年龄:</td>
                        <td><input type="text" name="studentAge"/></td>
                    </tr>
                    <tr>
                        <td>体重:</td>
                        <td><input type="text" name="studentWeight"/></td>
                    </tr>
                    <tr align="center">
                        <td colspan="2">
                            <input name="sure" type="submit" value="提 交"/>
                                
                            <input name="clear" type="reset" value="取 消"/>
                        </td>
                    </tr>
                </table>
            </form>
        </center>
    </body>
</html>
```

在图 8-29 中输入数据后单击"提交"按钮，将数据提交到 AddStudentServlet，即提交到 Servlet 文件处理数据，并对下一步操作进行处理，如图 8-30 所示。单击"确认"按钮返回系统主页面。

图 8-30　添加成功页面

**例 8-9**：addStudent.jsp 页面对应的控制器 Servlet。
AddStudentServlet.java

```java
package studentManage;
import java.io.IOException;
import javax.servlet.ServletException;
import javax.servlet.http.HttpServlet;
import javax.servlet.http.HttpServletRequest;
import javax.servlet.http.HttpServletResponse;
public class AddStudentServlet extends HttpServlet {
    protected void doGet(HttpServletRequest request, HttpServletResponse
            response) throws ServletException, IOException {
        String studentNumber=request.getParameter("studentNumber");
        String studentName=request.getParameter("studentName");
        String studentSex=request.getParameter("studentSex");
        String studentAge=request.getParameter("studentAge");
        String studentWeight=request.getParameter("studentWeight");
        DBJavaBean db=new DBJavaBean();
        if(db.addStudent(studentNumber,studentName,studentSex,
                studentAge,studentWeight)){
            response.sendRedirect("studentManage/message1.jsp");
        }else{
            response.sendRedirect("studentManage/addStudent.jsp");
        }
    }
    protected void doPost(HttpServletRequest request, HttpServletResponse
            response) throws ServletException, IOException {
        doGet(request, response);
    }
}
```

从例 8-9 代码中看出，该 Servlet（控制器）调用 DBJavaBean 类来处理添加学生信息的业务逻辑，即 DBJavaBean 封装处理 V（页面）的功能，这也是 MVC 设计模式的思想。本例子把对所有 V 的业务处理功能都封装到该 JavaBean 中了。在 MVC 设计模式的思想中，一个 V 对应一个处理 V 的 M（完成 V 功能的 JavaBean），V 提交到 C，C 获取 V 的数据并调用 M 在 C 中进行业务逻辑的处理，处理完后进行下一步的页面跳转，即添加成功页面调转到 message1.jsp，否则跳转到 addStudent.jsp。另外，使用 Servlet 文件需要在 web.xml 中配置，下面的 web.xml 文件提供了本实例用到的所有 Servlet 文件的配置。

例 8-10：处理添加学生信息页面的 DBJavaBean 类。

DBJavaBean.java

```java
package studentManage;
import java.sql.Connection;
import java.sql.DriverManager;
import java.sql.ResultSet;
import java.sql.Statement;
import javax.swing.JOptionPane;
public class DBJavaBean {
    private String driverName="com.mysql.jdbc.Driver";
    private String url=
    "jdbc:mysql://localhost:3306/student?useUnicode=
        true&characterEncoding=gbk";
    private String user="root";
    private String password="root";
    private Connection con=null;
    private Statement st=null;
    private ResultSet rs=null;
    public String getDriverName() {
        return driverName;
    }
    public void setDriverName(String driverName) {
        this.driverName = driverName;
    }
    public String getUrl() {
        return url;
    }
    public void setUrl(String url) {
        this.url = url;
    }
    public String getUser() {
        return user;
    }
    public void setUser(String user) {
        this.user = user;
    }
    public String getPassword() {
        return password;
    }
    public void setPassword(String password) {
```

```java
        this.password = password;
    }
    public Connection getCon() {
        return con;
    }
    public void setCon(Connection con) {
        this.con = con;
    }
    public Statement getSt() {
        return st;
    }
    public void setSt(Statement st) {
        this.st = st;
    }
    public ResultSet getRs() {
        return rs;
    }
    public void setRs(ResultSet rs) {
        this.rs = rs;
    }
//完成连接数据库操作,并生成容器返回
    public Statement getStatement(){
        try{
            Class.forName(getDriverName());
            con=DriverManager.getConnection(getUrl(),getUser(),getPassword());
            return con.createStatement();
        }catch(Exception e){
            e.printStackTrace();
            message("无法完成数据库的连接或者无法返回容器,请检查getStatement()方法!");
            return null;
        }
    }
//添加学生信息的方法
    public boolean addStudent(String studentNumber,String studentName,
        String studentSex,String studentAge,String studentWeight){
        try{
                String sql="insert into stuinfo"+
                    "(SID,SName,SSex,SAge,SWeight)"+"values("+"'"+
                    studentNumber+"'"+","+"'"+studentName+"'"+",
                    "+"'"+studentSex+"'"+","+"'"+studentAge+"'"+",
                    "+"'"+studentWeight+"'"+")";
                st=getStatement();
                int row=st.executeUpdate(sql);
                if(row==1){
                    st.close();
                    con.close();
                    return true;
                }else{
```

```java
                st.close();
                con.close();
                return false;
            }
        }catch(Exception e){
            e.printStackTrace();
            message("无法添加学生信息,请检查addStudent()方法!");
            return false;
        }
    }
    //查询所有学生信息,并返回 rs
    public ResultSet selectStudent(){
        try{
            String sql="select * from stuinfo";
            st=getStatement();
            return st.executeQuery(sql);
        }catch(Exception e){
            e.printStackTrace();
            message("无法查询学生信息,请检查aselectStudent()方法!");
            return null;
        }
    }
    //查询要修改的学生信息
    public ResultSet selectUpdateStudent(String NO){
        try{
            String sql="select * from stuinfo where SID='"+NO+"'";
            st=getStatement();
            return st.executeQuery(sql);
        }catch(Exception e){
            e.printStackTrace();
            message("无法查询到要修改学生的信息,请检查输入学生学号!");
            return null;
        }
    }
    //修改学生信息
    public boolean updateStudent(String studentNumber,String studentName,
        String studentSex,String studentAge,String studentWeight){
        try{
            String sql="update stuinfo set SID=
                '"+studentNumber+"',SName='"+studentName+"',
                SSex='"+studentSex+"',SAge='"+studentAge+"',
                SWeight='"+studentWeight+"'";
            st.executeUpdate(sql);
            return true;
        }catch(Exception e){
            e.printStackTrace();
            message("无法进行修改学生的信息,请检查updateStudent()方法!");
            return false;
```

```java
        }
    }
    //查询要删除的学生信息
    public ResultSet lookDeleteStudent(){
        try{
            String sql="select * from stuinfo";
            st=getStatement();
            return st.executeQuery(sql);
        }catch(Exception e){
            e.printStackTrace();
            message("无法查询到要删除学生的信息,请检LookDeleteStudent()方法!");
            return null;
        }
    }
    //查询要删除的学生信息
    public boolean DeleteStudent( String NO){
        try{
            String sql="delete  from stuinfo where SID="+NO;
            st=getStatement();
            st.executeUpdate(sql);
            return true;
        }catch(Exception e){
            e.printStackTrace();
            message("无法要删除学生的信息,请检DeleteStudent()方法!");
            return false;
        }
    }
    //一个带参数的信息提示框,供排错使用
    public void message(String msg){
        int type=JOptionPane.YES_NO_OPTION;
        String title="信息提示";
        JOptionPane.showMessageDialog(null,msg,title,type);
    }
}
```

例8-11:添加学生信息成功跳转的页面。

message1.jsp

```
<%@ page language="java" contentType="text/html; charset=UTF-8"
    pageEncoding="UTF-8"%>
<!DOCTYPE html PUBLIC "-//W3C//DTD HTML 4.01 Transitional//EN"
 "http://www.w3.org/TR/html4/loose.dtd">
<html>
<head>
<meta http-equiv="Content-Type" content="text/html; charset=UTF-8">
<title> message1</title>
</head>
<body bgcolor="CCCFFF">
    <br><br><br>
```

```
            <br><br><br>
            <center>
                <h3>添加成功! </h3>
                <form action="../studentManage/bottom.jsp">
                    <input type="submit" value="确 定">
            </center>
        </body>
</html>
```

**例 8-12**：在 web.xml 中配置 Servlet 文件。

web.xml

```
<?xml version="1.0" encoding="UTF-8"?>
<web-app version="3.0"
    xmlns=http://java.sun.com/xml/ns/javaee
    xmlns:xsi="http://www.w3.org/2001/XMLSchema-instance"
    xsi:schemaLocation="http://java.sun.com/xml/ns/javaee
    http://java.sun.com/xml/ns/javaee/web-app_3_0.xsd">
    <servlet>
        <servlet-name>AddStudentServlet</servlet-name>
        <servlet-class>studentManage.AddStudentServlet</servlet-class>
    </servlet>
    <servlet>
        <servlet-name>LookStudentServlet</servlet-name>
        <servlet-class>studentManage.LookStudentServlet
            </servlet-class>
    </servlet>
    <servlet>
        <servlet-name>UpdateStudentServlet</servlet-name>
        <servlet-class>studentManage.UpdateStudentServlet
            </servlet-class>
    </servlet>
    <servlet>
        <servlet-name>SelectUpdateStudentServlet</servlet-name>
        <servlet-class>studentManage.SelectUpdateStudentServlet
            </servlet-class>
    </servlet>
    <servlet>
        <servlet-name>DeleteStudentServlet</servlet-name>
        <servlet-class>studentManage.DeleteStudentServlet
            </servlet-class>
    </servlet>
    <servlet>
        <servlet-name>LookDeleteStudentServlet</servlet-name>
        <servlet-class>studentManage.LookDeleteStudentServlet
            </servlet-class>
    </servlet>
    <servlet-mapping>
        <servlet-name>AddStudentServlet</servlet-name>
```

```xml
        <url-pattern>/AddStudentServlet</url-pattern>
    </servlet-mapping>
    <servlet-mapping>
        <servlet-name>LookStudentServlet</servlet-name>
        <url-pattern>/LookStudentServlet</url-pattern>
    </servlet-mapping>
    <servlet-mapping>
        <servlet-name>UpdateStudentServlet</servlet-name>
        <url-pattern>/UpdateStudentServlet</url-pattern>
    </servlet-mapping>
    <servlet-mapping>
        <servlet-name>SelectUpdateStudentServlet</servlet-name>
        <url-pattern>/SelectUpdateStudentServlet</url-pattern>
    </servlet-mapping>
    <servlet-mapping>
        <servlet-name>DeleteStudentServlet</servlet-name>
        <url-pattern>/DeleteStudentServlet</url-pattern>
    </servlet-mapping>
    <servlet-mapping>
        <servlet-name>LookDeleteStudentServlet</servlet-name>
        <url-pattern>/LookDeleteStudentServlet</url-pattern>
    </servlet-mapping>
</web-app>
```

3. 学生信息查询功能的实现

单击图 8-28 的"学生信息查询"链接，出现如图 8-31 所示的页面（参考 left.jsp 中的"<a href="../LookStudentServlet" target="main">学生信息查询</a>"）。超链接到 LookStudentServlet 控制器（C）。

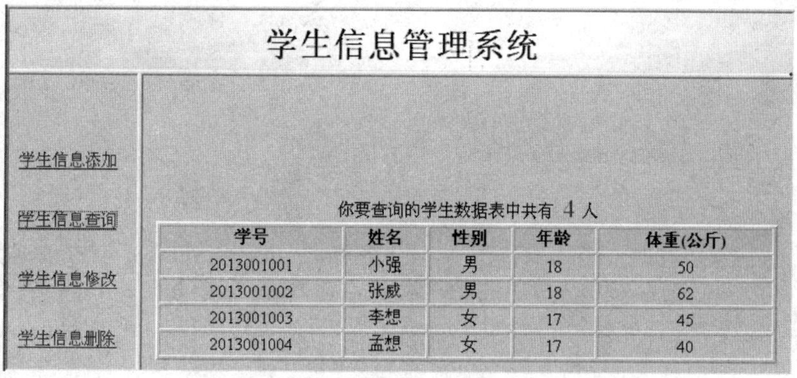

图 8-31 学生信息查询

例 8-13：LookStudentServlet 控制器。
LookStudentServlet.java

```
package studentManage;
import java.io.IOException;
import java.sql.ResultSet;
```

```java
import java.util.ArrayList;
import javax.servlet.ServletException;
import javax.servlet.http.HttpServlet;
import javax.servlet.http.HttpServletRequest;
import javax.servlet.http.HttpServletResponse;
import javax.servlet.http.HttpSession;
public class LookStudentServlet extends HttpServlet {
    protected void doGet(HttpServletRequest request, HttpServletResponse
            response) throws ServletException, IOException {
        try{
            DBJavaBean db=new DBJavaBean();
            ResultSet rs=db.selectStudent();
            //获取session对象
            HttpSession session=request.getSession();
            //声明一个集合对象保存数据
            ArrayList al=new ArrayList();
            while(rs.next()){
                //实例化学生对象用于把记录保存在对象中
                Student st=new Student();
                st.setStudentNumber(rs.getString("SID"));
                st.setStudentName(rs.getString("SName"));
                st.setStudentSex(rs.getString("SSex"));
                st.setStudentAge(rs.getString("SAge"));
                st.setStudentWeight(rs.getString("SWeight"));
                //把有数据的学生对象保存在集合中
                al.add(st);
                /*把集合对象保存在session中，以便于在lookStudent.jsp
                    中获取保存的数据*/
                session.setAttribute("al", al);
            }
            rs.close();
            response.sendRedirect("studentManage/lookStudent.jsp");
        }catch(Exception e){
            e.printStackTrace();
        }
    }
    protected void doPost(HttpServletRequest request, HttpServletResponse
            response) throws ServletException, IOException {
        doGet(request, response);
    }
}
```

例8-14：保存数据的Student类。

Student.java

```java
package studentManage;
public class Student {
    private String studentNumber;
    private String studentName;
```

```java
    private String studentSex;
    private String studentAge;
    private String studentWeight;
    public String getStudentNumber() {
        return studentNumber;
    }
    public void setStudentNumber(String studentNumber) {
        this.studentNumber = studentNumber;
    }
    public String getStudentName() {
        return studentName;
    }
    public void setStudentName(String studentName) {
        this.studentName = studentName;
    }
    public String getStudentSex() {
        return studentSex;
    }
    public void setStudentSex(String studentSex) {
        this.studentSex = studentSex;
    }
    public String getStudentAge() {
        return studentAge;
    }
    public void setStudentAge(String studentAge) {
        this.studentAge = studentAge;
    }
    public String getStudentWeight() {
        return studentWeight;
    }
    public void setStudentWeight(String studentWeight) {
        this.studentWeight = studentWeight;
    }
}
```

获取数据后页面跳转到 lookStudent.jsp 页面。

例 8-15：LookStudentServlet 控制器将页面跳转到 lookStudent.jsp。

lookStudent.jsp

```jsp
<%@page import="studentManage.Student"%>
<%@page import="java.util.ArrayList"%>
<%@page import="java.sql.*"%>
<%@ page language="java" contentType="text/html; charset=UTF-8"
    pageEncoding="UTF-8"%>
<!DOCTYPE html PUBLIC "-//W3C//DTD HTML 4.01 Transitional//EN"
 "http://www.w3.org/TR/html4/loose.dtd">
<html>
<head>
```

```jsp
<meta http-equiv="Content-Type" content="text/html; charset=UTF-8">
<title>学生信息查询</title>
</head>
    <body bgcolor="CCCFFF">
        <center>
            <br> <br> <br> <br> <br>
            <%
                //获取 al 中的数据，即集合中的数据
                ArrayList al=(ArrayList)session.getAttribute("al");
            %>
            你要查询的学生数据表中共有
            <font size="5" color="red">
                <%=al.size()%>
            </font>
            人
            <table border="2" bgcolor= "CCCEEE" width="600">
                <tr bgcolor="CCCCCC" align="center">
                    <th>学号</th>
                    <th>姓名</th>
                    <th>性别</th>
                    <th>年龄</th>
                    <th>体重(公斤)</th>
                </tr>
                <%
                    for(int i=0;i<al.size();i++){
                        Student st=(Student)al.get(i);
                %>
                <tr align="center">
                    <td><%=st.getStudentNumber()%> </td>
                    <td><%=st.getStudentName()%> </td>
                    <td><%=st.getStudentSex()%> </td>
                    <td><%=st.getStudentAge()%> </td>
                    <td><%=st.getStudentWeight()%> </td>
                </tr>
                <%
                    }
                %>
            </table>
        </center>
    </body>
</html>
```

4. 学生信息修改功能的实现

单击图 8-28 的"学生信息查询"链接，出现如图 8-32 所示的页面（参考 left.jsp 中的"<a href="lookUpdateStudent.jsp" target="main">学生信息修改</a>"）。超链接到 lookUpdateStudent.jsp 页面。

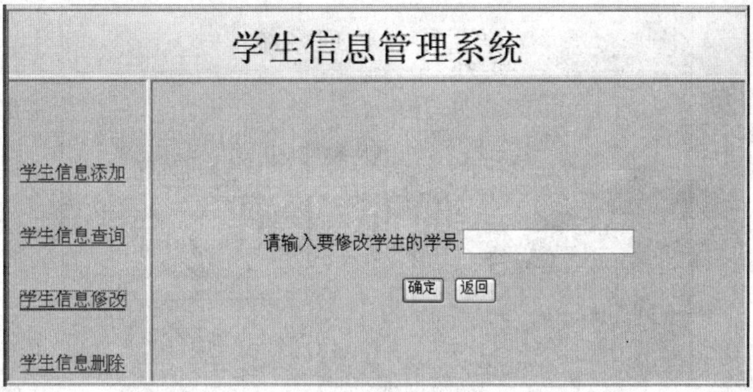

图 8-32 输入要修改学生的学号

**例 8-16**：输入要修改学生学号页面。

lookUpdateStudent.jsp

```jsp
<%@ page language="java" contentType="text/html; charset=UTF-8"
    pageEncoding="UTF-8"%>
<!DOCTYPE html PUBLIC "-//W3C//DTD HTML 4.01 Transitional//EN"
 "http://www.w3.org/TR/html4/loose.dtd">
<html>
<head>
<meta http-equiv="Content-Type" content="text/html; charset=UTF-8">
<title>学生信息修改</title>
</head>
    <body bgcolor="CCCFFF">
        <center>
            <br><br><br>
            <br><br><br>
            <form action="../SelectUpdateStudentServlet" method="post">
                <p>请输入要修改学生的学号：
                    <input type="text" name="studentNumber">
                </p>
                <p>
                    <input type="submit" value="确定"> 
                    <input type="button" value="返回"
                        onClick="javascript:history.go(-1)">
                </p>
            </form>
        </center>
    </body>
</html>
```

在图 8-32 中输入要修改学生信息的学号后单击"确定"按钮，将请求提交到 SelectUpdateStudentServlet 控制器进行处理，并将页面跳转到如图 8-33 所示的修改页面。

图 8-33　修改学生信息页面

**例 8-17**：lookUpdateStudent.jsp 对应的控制器。
SelectUpdateStudentServlet.java

```java
package studentManage;
import java.io.IOException;
import java.sql.ResultSet;
import java.util.ArrayList;
import javax.servlet.ServletException;
import javax.servlet.http.HttpServlet;
import javax.servlet.http.HttpServletRequest;
import javax.servlet.http.HttpServletResponse;
import javax.servlet.http.HttpSession;
public class SelectUpdateStudentServlet extends HttpServlet {
    protected void doGet(HttpServletRequest request, HttpServletResponse
            response) throws ServletException, IOException {
        try{
            DBJavaBean db=new DBJavaBean();
            String studentNumber=request.getParameter("studentNumber");
            ResultSet rs=db.selectUpdateStudent(studentNumber);
            HttpSession session=request.getSession();
            ArrayList al=new ArrayList();
            while(rs.next()){
                Student st=new Student();
                st.setStudentNumber(rs.getString("SID"));
                st.setStudentName(rs.getString("SName"));
                st.setStudentSex(rs.getString("SSex"));
                st.setStudentAge(rs.getString("SAge"));
                st.setStudentWeight(rs.getString("SWeight"));
                al.add(st);
                session.setAttribute("al",al);
            }
            rs.close();
            response.sendRedirect("studentManage/selectUpdateStudent.jsp");
        }catch(Exception e){
            e.printStackTrace();
```

```
        }
    }
    protected void doPost(HttpServletRequest request, HttpServletResponse
            response) throws ServletException, IOException {
        doGet(request, response);
    }
}
```

例 8-18：修改学生信息页面。

selectUpdateStudent.jsp

```
<%@page import="java.util.ArrayList"%>
<%@page import="studentManage.Student"%>
<%@ page language="java" contentType="text/html; charset=UTF-8"
    pageEncoding="UTF-8"%>
<!DOCTYPE html PUBLIC "-//W3C//DTD HTML 4.01 Transitional//EN"
 "http://www.w3.org/TR/html4/loose.dtd">
<html>
<head>
<meta http-equiv="Content-Type" content="text/html; charset=UTF-8">
<title>学生信息修改页面</title>
</head>
    <body bgcolor="CCCFFF">
        <center>
            <br> <br> <br>
            <h3>请修改学生信息！</h3>
            <form action="../UpdateStudentServlet">
            <table border="2" bgcolor= "CCCEEE" width="600">
                <%
                    ArrayList al=(ArrayList)session.getAttribute("al");
                    for(int i=0;i<al.size();i++){
                        Student st=(Student)al.get(i);
                %>
                <tr>
                    <td>学号:</td>
                    <td>
                        <input type="text" name="studentNumber"
                            value="<%=st.getStudentNumber()%>"/>
                    </td>
                </tr>
                <tr>
                    <td>姓名:</td>
                    <td>
                        <input type="text" name="studentName"
                            value="<%=st.getStudentName()%>"/>
                    </td>
                </tr>
                <tr>
                    <td>性别:</td>
```

```
                <td>
                    <input type="text" name="studentSex"
                        value="<%=st.getStudentSex()%>"/>
                </td>
            </tr>
            <tr>
                <td>年龄:</td>
                <td>
                    <input type="text" name="studentAge"
                        value="<%=st.getStudentAge()%>"/>
                </td>
            </tr>
            <tr>
                <td>体重:</td>
                <td>
                    <input type="text" name="studentWeight"
                        value="<%=st.getStudentWeight()%>"/>
                </td>
            </tr>
            <tr align="center">
                <td colspan="2">
                    <input name="sure" type="submit" value="修 改"/>
                        
                    <input name="clear" type="reset" value="取 消"/>
                </td>
            </tr>
        <%
            }
        %>
        </table>
    </center>
    </body>
</html>
```

在图 8-33 中对信息进行修改后单击"修改"按钮,将请求提交到 UpdateStudentServlet 控制器。

**例 8-19**:修改学生信息页面对应的控制器。

UpdateStudentServlet.java

```
package studentManage;
import java.io.IOException;
import javax.servlet.ServletException;
import javax.servlet.http.HttpServlet;
import javax.servlet.http.HttpServletRequest;
import javax.servlet.http.HttpServletResponse;
public class UpdateStudentServlet extends HttpServlet {
    protected void doGet(HttpServletRequest request, HttpServletResponse
        response) throws ServletException, IOException {
```

```java
            String studentNumber=request.getParameter("studentNumber");
            String studentName=request.getParameter("studentName");
            String studentSex=request.getParameter("studentSex");
            String studentAge=request.getParameter("studentAge");
            String studentWeight=request.getParameter("studentWeight");
            DBJavaBean db=new DBJavaBean();
            if(db.updateStudent(studentNumber,studentName,studentSex,
                studentAge,studentWeight)){
                response.sendRedirect("studentManage/message2.jsp");
            }else{
                response.sendRedirect("studentManage/lookUpdateStudent.jsp");
            }
        }
        protected void doPost(HttpServletRequest request, HttpServletResponse
                response)  throws ServletException, IOException {
            doGet(request, response);
        }
    }
```

修改成功后页面跳转到 message2.jsp，否则跳转到 lookUpdateStudent.jsp。

**例 8-20**：修改成功页面。

message2.jsp

```jsp
<%@ page language="java" contentType="text/html; charset=UTF-8"
    pageEncoding="UTF-8"%>
<!DOCTYPE html PUBLIC "-//W3C//DTD HTML 4.01 Transitional//EN"
 "http://www.w3.org/TR/html4/loose.dtd">
<html>
<head>
<meta http-equiv="Content-Type" content="text/html; charset=UTF-8">
<title>修改成功页面</title>
</head>
    <body bgcolor="CCCFFF">
        <br><br><br>
        <center>
            <h3>修改成功！</h3>
            <form action="../studentManage/bottom.jsp">
                <input type="submit" value="确 定">
        </center>
    </body>
</html>
```

**5. 学生信息删除功能的实现**

单击图 8-28 中的"学生信息删除"链接，出现如图 8-34 所示的页面（参考 left.jsp 中的"<a href="../LookDeleteStudentServlet" target="main">学生信息删除</a>"）。超链接到 LookDeleteStudentServlet 控制器。

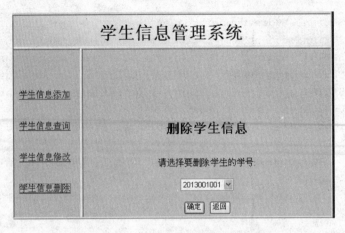

图 8-34 选择要删除学生的学号

例 8-21：选择删除的控制器。
LookDeleteStudentServlet.java

```java
package studentManage;
import java.io.IOException;
import java.sql.ResultSet;
import java.util.ArrayList;
import javax.servlet.ServletException;
import javax.servlet.http.HttpServlet;
import javax.servlet.http.HttpServletRequest;
import javax.servlet.http.HttpServletResponse;
import javax.servlet.http.HttpSession;
public class LookDeleteStudentServlet extends HttpServlet {
    protected void doGet(HttpServletRequest request, HttpServletResponse
        response) throws ServletException, IOException {
        try{
            DBJavaBean db=new DBJavaBean();
            ResultSet rs=db.lookDeleteStudent();
            HttpSession session=request.getSession();
            ArrayList al=new ArrayList();
            while(rs.next()){
                Student st=new Student();
                st.setStudentNumber(rs.getString("SID"));
                al.add(st);
                session.setAttribute("al", al);
            }
            rs.close();
            response.sendRedirect("studentManage/lookDeleteStudent.jsp");
        }catch(Exception e){
            e.printStackTrace();
        }
    }
    protected void doPost(HttpServletRequest request, HttpServletResponse
        response) throws ServletException, IOException {
```

```
            doGet(request, response);
    }
}
```

LookDeleteStudentServlet 控制器处理数据后页面跳转到 lookDeleteStudent.jsp。

**例 8-22**：选择删除的控制器跳转的页面。

lookDeleteStudent.jsp

```jsp
<%@page import="studentManage.Student"%>
<%@page import="java.util.ArrayList"%>
<%@page import="java.sql.*"%>
<%@ page language="java" contentType="text/html; charset=UTF-8"
    pageEncoding="UTF-8"%>
<!DOCTYPE html PUBLIC "-//W3C//DTD HTML 4.01 Transitional//EN"
 "http://www.w3.org/TR/html4/loose.dtd">
<html>
<head>
<meta http-equiv="Content-Type" content="text/html; charset=UTF-8">
<title>学生信息删除页面</title>
</head>
    <body bgcolor="CCCFFF">
        <center>
            <br><br><br>
            <br><br><br>
            <h2>删除学生信息</h2><br>
            <%
                ArrayList al=(ArrayList)session.getAttribute("al");
            %>
            <form action="../DeleteStudentServlet" method="post">
                <p>请选择要删除学生的学号:</p>
                <select name="NO">
                    <%
                        for(int i=0;i<al.size();i++){
                            Student st=(Student)al.get(i);
                    %>
                    <option value="<%=st.getStudentNumber()%>">
                        <%=st.getStudentNumber()%>
                    </option>
                    <%
                        }
                    %>
                </select>
                <p>
                    <input type="submit" value="确定"> 
                    <input type="button" value="返回"
                        onClick="javascript:history.go(-1)">
                </p>
            </form>
        </center>
    </body>
</html>
```

在图8-34中选择要删除学生的学号后单击"确定"按钮，将请求提交到DeleteStudentServlet控制器。

**例8-23**：删除控制器。

DeleteStudentServlet.java

```java
package studentManage;
import java.io.IOException;
import javax.servlet.ServletException;
import javax.servlet.http.HttpServlet;
import javax.servlet.http.HttpServletRequest;
import javax.servlet.http.HttpServletResponse;
public class DeleteStudentServlet extends HttpServlet {
    protected void doGet(HttpServletRequest request, HttpServletResponse
            response) throws ServletException, IOException {
        DBJavaBean db=new DBJavaBean();
        String NO=request.getParameter("NO");
        if(db.DeleteStudent(NO))
            response.sendRedirect("studentManage/message3.jsp");
    }
    protected void doPost(HttpServletRequest request, HttpServletResponse
            response) throws ServletException, IOException {
        doGet(request, response);
    }
}
```

删除后页面跳转到 message3.jsp。

**例8-24**：删除控制器跳转的页面。

message3.jsp

```jsp
<%@ page language="java" contentType="text/html; charset=UTF-8"
    pageEncoding="UTF-8"%>
<!DOCTYPE html PUBLIC "-//W3C//DTD HTML 4.01 Transitional//EN"
 "http://www.w3.org/TR/html4/loose.dtd">
<html>
<head>
<meta http-equiv="Content-Type" content="text/html; charset=UTF-8">
<title>删除成功页面</title>
</head>
    <body bgcolor="CCCFFF">
        <br><br><br>
        <center>
            <h3>删除成功! </h3>
            <form action="../studentManage/bottom.jsp">
                <input type="submit" value="确 定">
        </center>
    </body>
</html>
```

## 8.4　JSP 在数据库应用中的其他相关问题

### 8.4.1　分页技术

在实际应用中，如果从数据库中查询得到的记录特别多，超过了显示器屏幕范围，这时可将结果分页显示。本例使用的数据库以及表是上一节学生管理系统使用的数据库和表。

假设总记录数为 intRowCount，每页显示记录的数量为 intPageSize，总页数为 intPageCount，那么总页数的计算公式如下：

如果(intRowCount % intPageSize)>0，则 intPageCount= intRowCount / intPageSize+1。

如果(intRowCount % intPageSize)=0，则 intPageCount= intRowCount / intPageSize。

翻页后显示第 intPage 页的内容，将记录指针移动到(intPage-1)*intPageSize+1。

**例 8-25**：分页显示实例。

pageBreak.jsp

```jsp
<%@page import="java.sql.*"%>
<%@ page language="java" contentType="text/html; charset=UTF-8"
    pageEncoding="UTF-8"%>
<!DOCTYPE html PUBLIC "-//W3C//DTD HTML 4.01 Transitional//EN"
 "http://www.w3.org/TR/html4/loose.dtd">
<html>
<head>
<meta http-equiv="Content-Type" content="text/html; charset=UTF-8">
<title>分页实例</title>
</head>
    <body bgcolor="CCBBDD">
        <center>
            分页显示记录内容
            <hr>
            <table border="1" width="50%" bgcolor="cccfff" align="center">
                <tr>
                    <th>学号</th>
                    <th>姓名</th>
                    <th>性别</th>
                    <th>年龄</th>
                    <th>体重</th>
                </tr>
                <%
                    Class.forName("com.mysql.jdbc.Driver");
                    String url=
                        "jdbc:mysql://localhost:3306/student?useUnicode=true&characterEncoding=gbk";
                    String user="root";
                    String password="root";
                    Connection conn=DriverManager.getConnection(
                            url,user,password);
                    int intPageSize;   //一页显示的记录数
                    int intRowCount;   //记录总数
```

```jsp
        int intPageCount;    //总页数
        int intPage;         //待显示页码
        String strPage;
        int i;
        intPageSize =2;      //设置一页显示的记录数
        strPage = request.getParameter("page"); //取得待显示页码
        if(strPage==null){
    //表明在QueryString中没有page这一个参数,此时显示第一页数据
            intPage = 1;
        } else{
            //将字符串转换成整型
            intPage = java.lang.Integer.parseInt(strPage);
            if(intPage<1)
                intPage=1;
        }
        Statement stmt=conn.createStatement(
                ResultSet.TYPE_SCROLL_SENSITIVE,
                ResultSet.CONCUR_READ_ONLY);
        String sql="select * from stuinfo";
        ResultSet  rs=stmt.executeQuery(sql);
        rs.last(); //光标指向查询结果集中最后一条记录
        intRowCount = rs.getRow();   //获取记录总数
        intPageCount = (intRowCount+intPageSize-1) / intPageSize;
        //记算总页数
        if(intPage>intPageCount)
            intPage = intPageCount;//调整待显示的页码
        if(intPageCount>0){
            rs.absolute((intPage-1)*intPageSize + 1);
            //将记录指针定位到待显示页的第一条记录上
            //显示数据
            i=0;
            while(i<intPageSize && !rs.isAfterLast()){
%>
<tr>
    <td><%=rs.getString("SID")%> </td>
    <td><%=rs.getString("SName")%> </td>
    <td><%=rs.getString("SSex")%> </td>
    <td><%=rs.getString("SAge")%> </td>
    <td><%=rs.getString("SWeight")%> </td>
</tr>
<%
    rs.next();
    i++;
            }
        }
    %>
</table>
<hr>
<div align="center">
    第<%=intPage%>页 共<%=intPageCount%>页
    <%
        if(intPage<intPageCount){
```

```
                %>
                <a href="pageBreak.jsp?page=<%=intPage+1%>">下一页</a>
                <%
                }
                if(intPage>1){
                %>
                <a href="pageBreak.jsp?page=<%=intPage-1%>">上一页</a>
                <%
                }
                rs.close();
                stmt.close();
                conn.close();
                %>
            </div>
        </center>
    </body>
</html>
```

pageBreak.jsp 运行效果如图 8-35。

图 8-35　pageBreak.jsp 运行效果

### 8.4.2　常见中文乱码处理方式

在使用 MySQL 数据库时出现中文乱码，可以用以下几种方式进行解决。

1. 安装 MySQL 时设置编码方式

在安装 MySQL 时设置编码方式，设置为 gb2312 或者 tuf8，如图 8-36 所示。

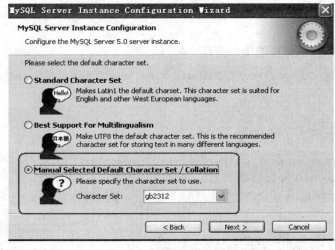

图 8-36　安装 MySQL 时设置编码方式

2. 创建数据库时设置字符集和整理

在创建数据库时将"字符集"和"整理"都设置为 gb2312，如图 8-37 所示。

图 8-37 设置数据库的编码方式

3. 创建数据表时设置字符集和整理

创建表时如果该字段需要输入中文，也需要把该字段的"字符集"和"整理"设置为 gb2312，如图 8-38 所示。

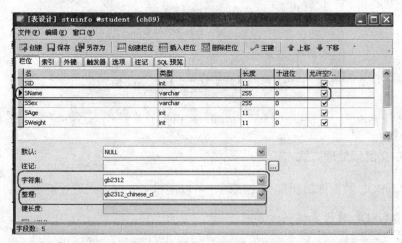

图 8-38 设置数据表的编码方式

4. 传送参数

通过传送参数设置数据库的编码方式。

```
jdbc:mysql://localhost/student? ?useUnicode=true&characterEncoding=gbk.
```

5. 代码转换

把中文转换为标准的字符方式。

```
String name=request.getParameter("StudentName");
Byte b[]=name.getBytes("ISO-8859-1");
name=new String(b,"UTF-8");
```

或者

```
String name=new
String(request.getParameter("StudentName ").getBytes("ISO-8859-1"),"UTF-8");
```

备注：在使用 MySQL 数据库时应从 1～5 开始。有时不需要使用代码转换时使用后反而会出现乱码问题。另外，在实现登录功能时，如果使用的是 SQL Server 数据库，数据库中的数据和输入的数据明明一致，但还是提示用户名和密码不对或者是无法登录，其中一个原因是在设计表时字段过长，这样数据后有空格，所以操作表时需要去掉空格，或者输入数据时后面加空格，一般建议去掉空格。

## 8.5 小　　结

本章主要介绍了在 JSP 项目开发中用到的 JDBC 技术，通过本章的学习应熟练使用 JDBC 来开发 JSP 项目。另外，本章通过一个 MVC 项目，综合运用了前 7 章的知识，达到巩固所学知识的效果，并为后面的基于 MVC 的 BBS 项目奠定基础。

## 8.6 习　　题

1. 简述 JDBC 以及其作用。
2. 简述 JDBC 驱动器的实现方式。
3. 简述 JDBC 连接数据库的步骤。
4. 把本章的分页技术写在一个 JavaBean 中，并在学生管理系统中应用。
5. 编写一个小型的用户管理系统，它可以查看所有用户的信息，包括姓名、地址等信息；能够对每个用户的个人信息进行修改；能够添加一个新用户或删除一个用户；还能够根据用户名查询用户的信息。
6. 按照 MVC 的思想开发一个宿舍管理系统，用来管理宿舍的日常活动。

# 第 9 章  JSP 标准标签库

本章主要介绍 JSP 标准标签库的使用方法，对 JSP 标准标签库的核心标签库、国际化标签库、XML 标签库、数据库标签库和函数标签库进行讲解，在讲解过程中结合具体示例演示 JSP 标准标签库在 JSP 中的使用方法。

## 9.1 JSP 标准标签库简介

### 9.1.1 概述

利用 JSP 的程序片可以编写出复杂的动态页面，但是过量程序片的使用、将大量 Java 代码嵌入 JSP 页面中，使得 Web 应用的开发效率大大降低，并且使系统的业务层和显示层混合在一起，对系统整体的架构产生不良影响。

JSP 标准标签库（JavaServer Pages Standard Tag Library，JSTL）是 Sun 公司在 Java EE 规范中提出的 JSP 标准标签库。利用 JSTL 能够实现 Web 应用程序中常用的功能，如条件判断、循环处理、数据格式化、XML 操作以及数据库访问等。通过使用 JSTL 能够大大减少 JSP 页面中程序片的数量，有助于提高 JSP 页面的开发效率，还为页面设计人员和程序开发人员进行分工合作提供了便利。

本章将就 JSTL 中的常用标签进行讲解。

### 9.1.2 JSTL 的使用

使用 JSTL 需要用到 jstl.jar 和 standard.jar 两个文件，需要将上述两个文件复制到 Web 应用程序的 WEB-INF/lib 目录中。可以从 http://tomcat.apache.org/taglibs/standard/获取上述 jar 文件。

为了在 JSP 页面中使用 JSTL，需要使用 taglib 指令，该指令声明当前 JSP 文件使用了某种标签库，同时声明了 JSP 中调用标签库的前缀。如在 JSP 页面中使用 JSTL 的核心库需要采用下列方式：

```
<%@ taglib prefix="c" uri=" http://java.sun.com/jsp/jstl/core" %>
```

其中，uri 属性指明了当前引用的标签库是 JSTL 中的核心库；prefix 属性指明了在 JSP 页面中如果需要调用核心库的标签需要用的前缀名，例如使用核心库的 out 标签进行输出操作，应采用下列方式：

```
<c:out value="${username }" default="anonymous" />
```

JSTL 包括 5 个不同分工的标签库，各标签库的信息如表 9-1 所示。

**例 9-1**：利用 JSPL 显示页面对象值。

example_9_1.jsp

```
<%@ page language="java" import="java.util.*" pageEncoding="UTF-8"%>
<%@ taglib prefix="c" uri="http://java.sun.com/jsp/jstl/core" %>
```

```
<!DOCTYPE HTML PUBLIC "-//W3C//DTD HTML 4.01 Transitional//EN">
<html>
  <body>
    <%
    String username = "Tom";
    pageContext.setAttribute("username",username);
    %>
    <c:out value="${username}" default="anonymous"></c:out>
  </body>
</html>
```

表 9-1 标签库相关信息表

| 标 签 库 | 建议前缀 | URI | 说 明 |
|---|---|---|---|
| 核心库 | c | http://java.sun.com/jsp/jstl/core | 条件判断、循环处理等基本功能 |
| 国际化库 | fmt | http://java.sun.com/jsp/jstl/fmt | 数据格式化、国际化 |
| 数据库标签库 | sql | http://java.sun.com/jsp/jstl/fmt | 数据库操作相关 |
| XML 标签库 | x | http://java.sun.com/jsp/jstl/xml | 操作 XML |
| 函数库 | fn | http://java.sun.com/jsp/jstl/functions | 常用函数库 |

在上述代码中，第 2 行利用 taglib 指令引入了 JSTL 的核心标签库。随后的程序片中声明了对象 username，并将其存入当前页面，之后利用 JSTL 核心库的 out 标签，将页面中的对象值输出。发布之后该示例的运行如图 9-1 所示。

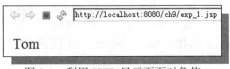

图 9-1 利用 JSTL 显示页面对象值

## 9.2 核心标签库

核心标签库中的标签主要实现变量控制、流程控制、页面跳转等功能。使用核心标签库时需要使用的 tablib 指令如下：

```
<%@ taglib prefix="c" uri="http://java.sun.com/jsp/jstl/core" %>
```

核心标签库中的常用标签归纳如表 9-2 所示。

表 9-2 核心标签库

| 标 签 | 说 明 |
|---|---|
| out | 计算表达式并进行输出 |
| set | 设置变量 |
| remove | 移除变量 |
| if | if 流程控制 |
| choose | |
| when | choose 标签和 when 和 otherwise 标签配合使用达到流程控制功能 |
| otherwise | |
| forEach | 实现遍历功能 |

续表

| 标　签 | 说　明 |
|---|---|
| forTokens | 分割字符串 |
| param | 添加一个参数，可以和 import、redirect 和 url 标签配合使用 |
| import | 包含一个外部资源 |
| redirect | 将客户端请求重定向到其他资源 |
| url | 动态地生成一个网址 |
| catch | 提供 JSP 页面的异常处理机制 |

- out 标签。该标签能够对表达式进行计算，并将结果输出到页面中。例如：

```
<c:out value="${ 3.14 * 10 } "> </c:out>
```

可以在其中指定 default 属性，当指定了 default 属性后，如果 value 属性的值为 null 时，会输出 default 属性的值。例如：

```
<c:out value="${ username }" default="anonymous"> </c:out>
```

如果当前上下文不存在 username 变量，则会在页面中输出 anonymous。

- set 标签。set 标签可以设置一个在特定范围内有效的变量值。例如：

```
<c:set var = "username" value="Tom"/>
```

上面语句的作用是设置了一个名为 username 的变量，该变量的值为 Tom，该变量的作用域为当前页面。也可以指定该变量的作用域为其他，例如：

```
<c:set var = "username" value="Tom" scope="session"/>
```

上面语句设置的 username 的作用域为 session。在设置变量值的时候还可以使用 EL 表达式的形式，例如：

```
<c:set var="username" value="${who.name}" scope="session"></c:set>
```

- remove 标签。remove 标签能够将特定作用域的某一变量值移除。例如：

```
<c:remove var="username" scope="session"/>
```

上面语句的作用是删除 session 作用域下的名为 username 的变量值。

- if 标签。if 标签提供对给定条件进行判断的功能。例如：

```
<c:if test="${empty(username)}">
    没有声明 username 变量
</c:if>
```

在上面的例子中，if 标签通过判定 test 属性是否为真来决定是否显示 if 标签体中的内容。

- choose、when、otherwise 标签。这 3 个标签可以结合起来使用达到条件判断的作用，例如：

```
<c:set var="x" value="-1"></c:set>
<c:choose>
    <c:when test="${x gt 1}">
        x > 1
    </c:when>
    <c:when test="${x gt 2}">
```

```
            x > 2
        </c:when>
        <c:otherwise>
            <c:out value="${x}" />
        </c:otherwise>
    </c:choose>
```

这里 choose 标签的作用类似于 Java 中 switch 语句的作用，when 标签类似于 switch 语句中 case 的作用，而 otherwise 标签类似于 switch 语句中 default 的作用。

choose 标签表明进入一个分支判断，然后依次判定其中的 when 标签的 test 属性是否为真，如果为真则显示当前 when 标签中的内容，然后结束 choose 标签，否则进行下一个 when 标签的判断。如果所有的 when 标签都不成立，最后执行 otherwise 标签的内容。

其中 otherwise 标签不是必须存在的，如果所有的 when 都判断不成立，则 choose 标签不显示任何内容。

注意：这里 when 标签和 otherwise 标签必须同 choose 标签结合使用。

• forEach 标签。forEach 标签用于遍历集合，其支持很多种集合类型，包括 java.util.Collenction，java.util.Map，java.util.Iterator，java.util.Enumeration 等。

例 9-2：利用 JSTL 实现复选框功能。

exp_2_1.html

```
<html>
<head>
<meta http-equiv="Content-Type" content="text/html; charset=UTF-8">
</head>
<body>
    <h3>您喜欢的水果有</h3>
    <form action="exp_2_2.jsp">
        苹果<input type="checkbox" value="apple" name="fruit" />
        橘子<input type="checkbox" value="orange" name="fruit" />
        香蕉<input type="checkbox" value="banana" name="fruit" /><br />
            <input type="submit" value="提交" />
    </form>
</body>
</html>
```

exp_2_2.jsp

```
<%@ page language="java" contentType="text/html; charset=UTF-8"
    pageEncoding="UTF-8"%>
<%@ taglib prefix="c" uri="http://java.sun.com/jsp/jstl/core" %>
<html>
<body>
<c:forEach items="${paramValues.fruit}" var="f">
    ${ f }
</c:forEach>
</body>
</html>
```

运行效果如图 9-2、图 9-3 所示。

图 9-2  input.html 运行效果

图 9-3  output.jsp 运行效果

同时，forEach 标签还支持对一组由逗号分隔的字符串进行遍历。例如：

```
<c:forEach items="1,2,3,4,5" var="num">
    ${num }
</c:forEach>
```

将会输出 1 2 3 4 5 的字符串序列。

注意：forEach 标签分隔字符串时，只可以使用英文逗号进行分隔，如果需要灵活定制分隔符号，则应使用 forTokens 标签。

• forTokens 标签。forTokens 标签可以将给定的字符串按照特定的分隔符进行分隔。分隔符可以同时指定一个字符或多个字符，如果指定多个字符作为分隔符，则只要含有分隔符之一的位置都会进行分隔。分隔符通过 forToken 标签的 delims 属性指定。例如：

```
<c:forTokens items="1_2,3,_4_,5" var="num" delims="_,">
    ${num }<br/>
</c:forTokens>
```

运行的效果如图 9-4 所示。

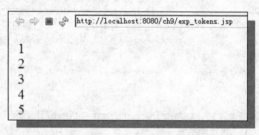

图 9-4  运行效果图

• import 标签。import 标签的作用类似于<jsp:include>指令，能够将一个资源包含入当前页面。例如：

```
<c:import url="a.txt"></c:import>
```

在包含入动态资源时，还可以通过在 import 标签中添加 param 子标签实现向其传递参数。例如：

```
<c:import url="3.jsp">
    <c:param name="name" value="Tom"></c:param>
```

```
    <c:param name="age" value="19"></c:param>
</c:import>
```

- redirect 标签。redirect 标签可以实现客户端请求的重定向功能，类似 import 标签该标签也可以嵌入 param 子标签实现向重定向的页面传递参数。例如：

```
<c:redirect url="login.jsp">
    <c:param name="username" value="Tom"></c:param>
</c:redirect>
```

- url 标签。url 标签的作用是动态地生成一个 URL 地址。例如：

```
<c:url value="http://www.sina.com.cn"></c:url>
```

类似于 import 标签，url 标签也能够嵌入 param 子标签传递参数。例如：

```
<c:url value="http://www.sina.com.cn">
    <c:param name="name" value="Tom"></c:param>
</c:url>
```

上面例子运行产生的 URL 地址为 http://www.sina.com.cn?name=Tom。

- catch 标签。catch 标签提供 JSP 页面的异常处理机制。将可能发生异常的程序片写在 catch 标签之间，则如果出现异常，这些代码将终止执行，但不会影响到整个页面的运行。例如：

```
<c:catch var="errMsg">
    <%
        int x = 10;
        int y = 0;
        int z = x/y;
    %>
</c:catch>
<c:out value="${errMsg}"></c:out>
```

上面例子运行的结果为在页面中显示出下列语句：

```
java.lang.ArithmeticException: / by zero
```

## 9.3 国际化标签库

国际化标签库中的标签主要负责实现区域设置、创建本地信息、格式化数字、格式化货币、格式化日期和时间等功能。使用核心标签库时需要使用的 tablib 指令如下：

```
<%@ taglib prefix="fmt" uri="http://java.sun.com/jsp/jstl/fmt" %>
```

国际化标签库中的标签归纳如表 9-3 所示。

- setLocale 标签。setLocal 标签用来设定语言地区代码。例如：

```
<fmt:setLocale value="zh_CN" scope="session"/>
```

上面将用户会话的上下文环境设置成为了中文。其中 value 属性的值用来指定地区代码；scope 的值用来指定上下文信息的作用域，该值可以取 page、request、session 或 application，分别对象相应的作用域。

注意：通过 setLocal 标签设置本地上下文信息后，浏览器中设置的语言首选项将会失效。

表 9-3 国际化标签库

| 标　　签 | 说　　明 |
| --- | --- |
| setLocale | 设定语言地区 |
| requestEncoding | 设定字符串编码 |
| setBundle | 用来实现文字的国际化显示 |
| bundle | |
| message | |
| param | |
| formatNumber | 本地化处理数字信息 |
| parseNumber | |
| formatDate | 本地化处理日期信息 |
| parseDate | |
| setTimeZone | 设置本地的上下文环境 |
| timeZone | |

- requestEncoding 标签。requestEncoding 标签用来设定字符串的编码方式，其作用类似于 ServletRequest.setCharacterEncoding()方法。例如：

```
<fmt:requestEncoding value="UTF-8"/>
```

相当于执行了语句

```
request.setCharacterEncoding("UTF-8");
```

- setBundle、bundle、message、param 标签。这些标签用来显示特定语言环境下的文本信息，实现文本显示的国际化。

国际化需要使用到资源文件，一般来说，资源文件是以.properties 为后缀的一个文本文件，其中的内容是以 Key = Value 方式存放的一系列信息映射。例如一个名为 xx_zh_CN.properties 的资源文件如下：

```
xx_zh_CN.properties
title = \u6807\u9898
now = \u73b0\u5728\u65f6\u95f4 : {0}
```

其中以"\u"开头的一系列十六进制数字表示一些本地字符的 Unicode 转换码，例如，\u6807\u9898 是"标题"这两个字的 Unicode 码。\u73b0\u5728\u65f6\u95f4 则是"现在时间" 4 个字的 Unicode 表示。如果需要转换字符串到 Unicode 表示，可以使用 JDK 安装目录下 bin 文件夹中的 native2ascii.exe 进行处理；或者使用 MyEclipse 的内置资源编辑器进行处理。

setBundle 标签用来设置一个特定作用域内有效的默认资源文件，供 message 标签使用。

bundle 标签设置一个资源文件供嵌套在其内部的 message 标签使用。

message 标签从指定的资源包中取出指定关键字进行显示。

对于待显示的国际化信息，如果存在占位符（形式如"{0}、{1}"），则可以在 message 标签中通过使用 param 子标签进行动态填充。

下面通过一个示例说明这些标签的使用方法。

例 9-3：国际化标签使用。

首先在网站的 WEB-INF/classes 目录下建立资源文件 message_zh_CN.properties，内容如下：

```
info= \u6b22\u8fce\u8bbf\u95ee{0}
now=\u73b0\u5728\u65f6\u95f4 : {0}
```

然后建立页面 exp_3.jsp，内容如下：

```
<%@ page language="java" contentType="text/html; charset=UTF-8"
    pageEncoding="UTF-8" import="java.util.Date"%>
<%@ taglib prefix="fmt" uri="http://java.sun.com/jsp/jstl/fmt" %>
<html>
<head>
<meta http-equiv="Content-Type" content="text/html; charset=UTF-8">
</head>
<body>
<fmt:setBundle basename="message_zh_CN" scope="request"/>
<%
    Date currentTime = new java.util.Date();
    request.setAttribute("currentTime",currentTime);
%>
<fmt:message key="info">
    <fmt:param value="国际化页面"></fmt:param>
</fmt:message>
<br/>
```

原来显示时间：${currentTime}<br/>

```
<fmt:bundle basename="message_zh_CN">
    <fmt:message key="now">
        <fmt:param value="${currentTime}"></fmt:param>
    </fmt:message>
</fmt:bundle>
</body>
</html>
```

运行结果如图 9-5 所示。

图 9-5　运行效果图

- formatNumber 和 parseNumber 标签。formatNumber 标签用指定的上下文环境显示数字内容，可以显示为数字、货币或者百分数。例如：

```
<fmt:setLocale value="zh_CN"/>
<fmt:formatNumber value="31415926" type="number"></fmt:formatNumber><br/>
<fmt:formatNumber value="31415926" type="currency"></fmt:formatNumber><br/>
<fmt:formatNumber value="31415926" type="percent"></fmt:formatNumber><br/>
<fmt:setLocale value="en_US"/><br/>
<fmt:formatNumber value="31415926" type="number"></fmt:formatNumber><br/>
<fmt:formatNumber value="31415926" type="currency"></fmt:formatNumber><br/>
<fmt:formatNumber value="31415926" type="percent"></fmt:formatNumber>
```

最后的运行结果如下：

```
31,415,926
￥31,415,926.00
3,141,592,600%

31,415,926
$31,415,926.00
3,141,592,600%
```

其中 formatNumber 标签的 value 属性用来指定待格式化的数字，type 属性则指定格式化数字的类型，可以取 number、currency 或 percent 三种值。

paeseNumber 标签用来解析存储于其 value 属性中的值，并将结果作为 java.lang.Number 类的实例返回。例如：

```
<fmt:parseNumber value="3.14" type="number" var="num1"></fmt:parseNumber>
<fmt:parseNumber value="$3.14" type="currency" var="num2"></fmt:parseNumber>
<fmt:parseNumber value="3.14%" type="percent" var="num3"></fmt:parseNumber>
<c:out value="${num1}" /> <br/>
<c:out value="${num2}" /> <br/>
<c:out value="${num3}" />
```

最后的运行结果如下：

```
3.14
3.14
0.031400000000000004
```

注意：小数值的转换可能存在精度的问题。按照货币金额进行转换时，要注意当前计算机使用的语言，否则会出现无法转换。

- formatDate 和 parseDate 标签。formatDate 标签能够根据需要格式化指定的日期和时间。例如：

```
<%
  java.util.Date currentTime = new java.util.Date();
  request.setAttribute("currentTime",currentTime);
%>
<fmt:formatDate value="${currentTime}" pattern="yyyy-MM-dd HH:mm:ss" />
```

将显示出类似于"2013-04-12 16:40:58"的结果，其中 formatDate 标签中的 pattern 属性用来指定时间显示的格式。

parseDate 标签能够解析日期和时间。例如：

```
<fmt:parseDate value="2013_4_1_12_12_12" var="time"
pattern="yyyy_MM_dd_HH_mm_ss"></fmt:parseDate>
<fmt:formatDate value="${time}" pattern="yyyy-MM-dd HH:mm:ss" />
```

则最终运行的显示结果为："2013-04-01 12:12:12"。其中 parseDate 中的 value 属性用来指定待转换日期的字符串值，pattern 属性指定转换的格式，var 属性用来声明存储转换后的日期变量。

注意：对于 formatDate 和 parseDate 标签中的 pattern 属性的值应当是符合 java.text.Simple-

DateFormat 类约定的模式字符串。parseDate 标签要解析的字符串值必须严格符合 pattern 规定的样式。

## 9.4 数据库标签库

数据库标签库为快捷开发系统提供了数据库的访问方式，但为了系统的业务分层，建议在实际开发中采用 JavaBean 进行数据库访问。

使用数据库标签库，需要使用的 taglib 指令如下：

```
<%@ taglib prefix="sql" uri="http://java.sun.com/jsp/jstl/sql" %>
```

数据库标签库中的常用标签归纳如表 9-4 所示。

表 9-4　数据库标签库

| 标签 | 说明 |
| --- | --- |
| setDataSource | 配置数据源 |
| query | 对数据库进行查询操作 |
| update | 对数据库进行更改操作 |
| param | 为查询语句插入参数 |
| dateParam | 为查询语句插入日期或时间参数 |
| transaction | 对数据库进行事务处理 |

- sqlDataSource 标签。sqlDataSource 标签的作用是设置数据源。例如：

```
<sql:setDataSource
    driver="com.mysql.jdbc.Driver"
    user="root"
    password="root"
    url="jdbc:mysql://127.0.0.1/ch9"
    var="sds"
/>
```

其中 driver 属性对应于数据库的连接驱动，当前示例用的数据库是 MySql，所以需要在 Web 应用程序的 WEB-INF/lib 目录中添加 MySql 的驱动 jar 包。user 属性对应数据库的登录名。password 属性对应数据库的登录口令。url 属性对应数据库访问的 URL。var 属性为数据源的变量名。

sqlDataSource 标签还提供了 JNDI 的配置方式。例如：

```
<sql:setDataSource dataSource="jdbc/MyDB"/>
```

这里的 dataSource 属性对应的是 Java EE 应用服务器上的数据源 JNDI 名称。

- query 标签。query 标签实现查询数据库的功能。能够在 query 标签中使用预处理语句（基于 java.sql.PreparedStatement 接口），这时可以在 query 标签中嵌套使用子标签 param 或 dateParam 在为 SQL 语句设置参数。例如：

```
<sql:query var="rs1" sql="select * from t_user" dataSource="${sds}">
    </sql:query>
```

上面 query 标签中 sql 属性是执行的查询语句，dataSource 属性是当前查询采用的数据源。

在执行查询之后,结果集分配给 var 属性指定的变量,改返回值是 javax.servlet.jsp.jstl.sql.Result 接口的实例,该接口提供了一系列方法对结果集的属性进行访问,如表 9-5 所示。

表 9-5 javax.servlet.jsp.jstl.sql.Result 接口

| 属 性 | 说 明 |
| --- | --- |
| Rows | 一组 SortedMap 对象,对应结果集中的数据行 |
| rowsByIndex | 一系列数组,每个数组对应结果集的一行 |
| columnNames | 一组字符串,每个字符串对应结果集的一个列名,顺序与 rowsByIndex 一致 |
| rowCount | 结果集的行数 |
| limitedByMaxRows | 判断 query 是否设置了 maxRows 属性 |

可以在 query 标签中使用带参数的查询语句,例如:

```
<sql:query var="rs1" sql="select * from t_user where age > ?" dataSource="${sds}">
    <sql:param>21</sql:param>
</sql:query>
```

还可以在 query 标签中指定 maxRows 属性和 startRow 属性来定制返回结果集的最多行数和其实行号,例如:

```
<sql:query var="rs1" sql="select * from t_user where age > ?" dataSource=
        "${sds}" maxRows="3" startRow="0">
    <sql:param>10</sql:param>
</sql:query>
```

对于 dateParam 标签,和 param 标签的作用类似,但 dateParam 的 value 属性值必须为 java.util.Date 的实例,并且 dateParam 的 type 属性值必须为 date、time 或 timestamp,如果是时间用 time,如果是日期用 date,如果是日期和时间则使用 timestamp。

• update 标签。update 标签的作用是实现对数据库的插入、更新、删除操作。update 的使用方法和 query 标签完全类似。但 var 属性对应的变量为 java.lang.Integer 类型,表示的是数据库执行 update 操作后更新的记录数。例如:

```
<sql:update var="count" sql="update t_user set age=? where username=?"
        dataSource="${sds}">
    <sql:param value="32"></sql:param>
    <sql:param>李四</sql:param>
</sql:update>
```

影响的记录数:<c:out value="${count}"></c:out>

注意:param 标签的值可以写到其 value 属性中,也可以写到标签体中。

• transaction 标签。transaction 标签用来保证一系列数据库操作的事务完整性。在 transaction 标签内嵌套的子标签将被视为一个事务,它们共用 transaction 的 dataSource 属性。例如:

```
<sql:transaction dataSource="${sds}" isolation="read_committed">
    <sql:update sql = "insert into t_user(username,age) values(?,?)">
        <sql:param>Tom</sql:param>
        <sql:param>18</sql:param>
    </sql:update>
```

```
    <sql:update sql="update t_user set age=? where username='Tom'" >
        <sql:param>19</sql:param>
    </sql:update>
</sql:transaction>
```

例子中将两个 update 操作定制成了一个事务，其中 transaction 标签中的 isolation 属性用于指定事务处理的隔离级别，可以取的值有 read_uncommitted、read_committed、repeatable_read 和 serializable，越往后隔离级别越严格，对系统的并发性产生的影响越大。建议：采用 read_committed 级别就能满足一般系统需要。

注意：嵌套到 transaction 标签中的 query 或 update 子标签不能够设置自己的 dataSource 属性。

**例 9-4**：对数据的操作。

本示例采用的数据库的初始化语句如下：

```
SET FOREIGN_KEY_CHECKS=0;
CREATE TABLE 't_user' (
  'id' bigint(20) NOT NULL AUTO_INCREMENT,
  'username' varchar(50) DEFAULT NULL,
  'age' int(11) DEFAULT NULL,
  'department' varchar(50) DEFAULT NULL,
  PRIMARY KEY ('id')
) ENGINE=InnoDB AUTO_INCREMENT=7 DEFAULT CHARSET=utf8;
INSERT INTO 't_user' VALUES ('1', '张三', '20', '人力资源');
INSERT INTO 't_user' VALUES ('2', '李四', '20', '销售');
INSERT INTO 't_user' VALUES ('3', '王五', '21', '售后服务');
INSERT INTO 't_user' VALUES ('4', '赵六', '22', '市场营销');
INSERT INTO 't_user' VALUES ('5', '钱七', '20', '售后服务'););
```

建立好的数据表内容如图 9-6 所示。

| id | username | age | department |
|----|----------|-----|------------|
| 1 | 张三 | 20 | 人力资源 |
| 2 | 李四 | 20 | 销售 |
| 3 | 王五 | 21 | 售后服务 |
| 4 | 赵六 | 22 | 市场营销 |
| 5 | 钱七 | 20 | 售后服务 |

图 9-6　数据库表内容

**exp_9_4.jsp**

```
<%@ page language="java" import="java.util.*" pageEncoding="UTF-8"%>
<%@ taglib prefix="c" uri="http://java.sun.com/jsp/jstl/core" %>
<%@ taglib prefix="sql" uri="http://java.sun.com/jsp/jstl/sql" %>
<!DOCTYPE HTML PUBLIC "-//W3C//DTD HTML 4.01 Transitional//EN">
<html>
  <body>
<sql:setDataSource
    driver="com.mysql.jdbc.Driver"
    user="root"
    password="root!"
```

```
    url="jdbc:mysql://127.0.0.1/ch9"
    var="sds"
/>
<sql:query var="rs1" sql="select * from t_user "
        dataSource="${sds}"></sql:query>
数据库中的人员：<br/>
<c:forEach items="${rs1.rows}" var="row">
    <c:out value="${row.username}"></c:out><br/>
</c:forEach>
对"李四"进行更新操作：<br/>
<sql:update var="count" sql="update t_user set age=
        ? where username=?" dataSource="${sds}">
    <sql:param value="32"></sql:param>
    <sql:param>李四</sql:param>
</sql:update>
影响的记录数：<c:out value="${count}"></c:out><br/>
添加新的人员Tom，并更改Tom的年龄。
<sql:transaction dataSource="${sds}" isolation="read_committed">
    <sql:update sql = "insert into t_user(username,age) values(?,?)">
        <sql:param>Tom</sql:param>
        <sql:param>18</sql:param>
    </sql:update>
    <sql:update sql="update t_user set age=? where username='Tom'" >
        <sql:param>19</sql:param>
    </sql:update>
</sql:transaction>
  </body>
</html>
```

该程序的运行结果如图 9-7 所示。运行之后数据库的内容如图 9-8 所示。

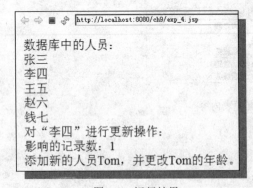

图 9-7　运行结果　　　　　　　图 9-8　运行之后数据库的内容

注意：需要在 Web 应用程序中引入数据库驱动，当前例子使用的数据库为 MySQL。

## 9.5　XML 标签库

JSTL 提供了对 XML 的灵活支持，能够方便地处理 XML 文件。使用 XML 标签库的 taglib 指令如下：

```
<%@ taglib prefix="x" uri="http://java.sun.com/jsp/jstl/xml" %>
```

另外，需要注意的是，对于本节的一些例子，可能需要使用到 Apache 的 xalan 项目，该项目的 jar 文件可从 http://archive.apache.org/dist/xml/xalan-j/网页获取。

XML 标签库中的常用标签归纳如表 9-6 所示。

表 9-6　XML 标签库

| 标　　签 | 说　　明 |
| --- | --- |
| parse | 用于解析 XML 文件 |
| out | 输出 XML 文档中的元素值 |
| set | 计算 XPath 表达式值，并保存到一个变量中 |
| if | 和 JSTL 核心库中的同名标签作用类似，进行流程控制 |
| choose | |
| when | |
| otherwise | |
| foreach | |
| transform | 将指定的 XML 转换成其他格式 |
| param | 用于 transform 中传递参数 |

- parse 标签。parse 标签用于解析 XML 文档，该 XML 文档可以写在 parse 标签体内，也可以结合 JSTL 核心库的 import 标签使用。例如：

```
<x:parse var="axml">
    <user>
        <name>Tom</name>
        <age>18</age>
    </user>
</x:parse>
```

或：

```
<c:import url="user.xml" var="xml_doc"></c:import>
<x:parse var="bxml" doc="${xml_doc}"></x:parse>
```

- set 标签。set 标签可以从 XML 文档中取出内容并存入变量中。例如：

```
<x:parse var="cxml">
    <user>
        <name>Rose</name>
        <age>20</age>
    </user>
</x:parse>
<x:set var="user" select="$cxml/user"/>
```

这时 user 变量中存的是 XML 的部分内容，如果要输出 Rose 的名字，则可以使用下面的语句：

```
<x:out select="$user/name"/>
```

- if、choose、when、otherwise、foreach 标签。这些标签的作用类似于 JSTL 核心库中的流程控制标签，主要负责条件判断和循环等功能。但这些标签中处理的对象是 XML 文件，它们通过 select 属性指定 XPath 表达式来进行条件判断，而 JSTL 核心库使用的是 EL 表达式。下面以示例讲解其用法。

**例 9-5**：流程控制示例。该示例可能需要用到 xalan.jar 文件（视调试环境而定），该文件可从 http://archive.apache.org/dist/xml/xalan-j/ 网页获取。

exp_5.jsp

```
<%@ page language="java" import="java.util.*" pageEncoding="UTF-8"%>
<%@ taglib prefix="x" uri="http://java.sun.com/jsp/jstl/xml" %>
<!DOCTYPE HTML PUBLIC "-//W3C//DTD HTML 4.01 Transitional//EN">
<html>
  <body>
<x:parse var="axml">
    <users>
        <user>
            <name>Tom</name>
            <age>18</age>
        </user>
        <user>
            <name>Mike</name>
            <age>19</age>
        </user>
        <user>
            <name>Rose</name>
            <age>20</age>
        </user>
    </users>
</x:parse>
<x:forEach select="$axml/users/user">
    <x:out select="name"/>,<x:out select="age"/>,
    <x:if select="age=19">刚好 19 岁</x:if>
    <x:choose>
        <x:when select="age<19">小于 19 岁</x:when>
        <x:when select="age>19">大于 19 岁</x:when>
    </x:choose>
    <br/>
</x:forEach>
  </body>
</html>
```

运行结果如图 9-9 所示。

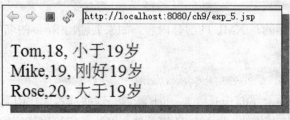

图 9-9 运行结果图

- transform 标签。XML 文件可以通过 XSL 样式表来进行格式处理。transform 标签可以将 XML 文档转换成指定的样式。

**例 9-6**：transform 示例。

该示例可能需要用到 serializer.jar 文件（视调试环境而定），该文件可从 http://archive.apache.org/dist/xml/xalan-j/ 网页获取。

exp_6.xsl

```xml
<?xml version="1.0" encoding="UTF-8"?>
<xsl:stylesheet version="1.0" xmlns:xsl="http://www.w3.org/1999/XSL/Transform">
    <xsl:template match="/">
        <html>
        <head>
        <title>hello</title>
        </head>
        <body>
        <xsl:for-each select="pub">
        <table border="1" cellspacing="0">
        <caption style="font-size: 150%; font-weight: bold">Hello</caption>
        <tr>
        <th>name</th>
        <th>sex</th>
        <th>hight</th>
        </tr>
        <xsl:for-each select="item">
        <tr>
        <td><xsl:value-of select="name"/></td>
        <td><xsl:value-of select="sex"/></td>
        <td><xsl:value-of select="hight"/></td>
        </tr>
        </xsl:for-each>
        </table>
        </xsl:for-each>
        </body>
        </html>
    </xsl:template>
</xsl:stylesheet>
```

exp_6.xml

```xml
<?xml version="1.0" encoding="UTF-8"?>
<pub>
<item>
<name>张三 </name>
<sex>男 </sex>
<hight>175</hight>
</item>
<item>
<name>李四 </name>
```

```
        <sex>女 </sex>
        <hight>169</hight>
    </item>
    <item>
    <name>王五 </name>
    <sex>男 </sex>
    <hight>180</hight>
    </item>
</pub>
```

exp_6.jsp

```
<%@ page language="java" contentType="text/html; charset=UTF-8"
    pageEncoding="UTF-8"%>
<%@ taglib prefix="x" uri="http://java.sun.com/jsp/jstl/xml" %>
<%@ taglib prefix="c" uri="http://java.sun.com/jsp/jstl/core" %>
<!DOCTYPE html PUBLIC "-//W3C//DTD HTML 4.01 Transitional//EN"
        "http://www.w3.org/TR/html4/loose.dtd">
<html>
<head>
<meta http-equiv="Content-Type" content="text/html; charset=UTF-8">
<title>Insert title here</title>
</head>
<body>
<c:import url=" exp_6.xsl" var="xsl" charEncoding="UTF-8"></c:import>
<c:import url=" exp_6.xml" var="xml" charEncoding="UTF-8"></c:import>
<x:transform doc="${xml }" xslt="${xsl }"></x:transform>
</body>
</html>
```

该示例的运行结果如图9-10所示。

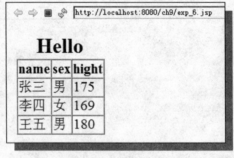

图 9-10 运行结果

注意：本实例中的 exp_6.xsl 和 exp_6.xml 文件所在的位置和 exp_6.jsp 在同一目录下。

## 9.6 函数标签库

JSTL 的函数标签库主要提供了用于字符串操作的方法。使用函数标签库的 taglib 指令如下：

```
<%@ taglib prefix="fn" uri="http://java.sun.com/jsp/jstl/functions" %>
```

JSTL 的函数标签库中的常用标签归纳如表 9-7 所示。

表 9-7 函数标签库

| 标　　签 | 说　　明 |
| --- | --- |
| contains | 判断字符串是否包含另外一个字符串 |
| containsIgnoreCase | 判断字符串是否包含另外一个字符串(大小写无关) |
| endsWith | 判断字符串是否以另外字符串结束 |
| escapeXml | 把一些字符转成 XML 表示，例如 "<" 字符应该转为&lt; |
| indexOf | 子字符串在母字符串中出现的位置 |
| join | 将数组中的数据联合成一个新字符串，并使用指定字符隔开 |
| length | 获取字符串的长度，或者数组的大小 |
| replace | 替换字符串中指定的字符 |
| split | 把字符串按照指定字符切分 |
| startsWith | 判断字符串是否以某个子串开始 |
| substring | 获取子串 |
| substringAfter | 获取从某个字符所在位置开始的子串 |
| substringBefore | 获取从开始到某个字符所在位置的子串 |
| toLowerCase | 转为小写 |
| toUpperCase | 转为大写字符 |
| trim | 去除字符串前后的空格 |

函数标签的使用方法都是一致的，其格式如下：

```
${fn:method(arg1…)}
```

即将所需使用的函数标签放置在 EL 表达式中即可计算出所需结果。

下面通过一个示例来说明函数标签库的使用方法。

例 9-7：函数标签库示例。

exp_7.jsp

```
<%@ page language="java" import="java.util.*" pageEncoding="UTF-8"%>
<%@ taglib prefix="c" uri="http://java.sun.com/jsp/jstl/core" %>
<%@ taglib prefix="fn" uri="http://java.sun.com/jsp/jstl/functions" %>
<!DOCTYPE HTML PUBLIC "-//W3C//DTD HTML 4.01 Transitional//EN">
<html>
  <body>
  <%
String[] arr = new String[]{"1","2","3","4","5"};
request.setAttribute("arr",arr);
 %>
  <c:set value="Hello World! <br>" var="str"></c:set>
初始的字符串数组为：arr = ["1","2","3","4","5"] <br/>
```

```
        初始的字符串 str ： <c:out value="${str}" default="nothing"/><br/>
        直接显示的效果为(&lt;br&gt;被解析为换行)：${str }<br/>
\${fn:contains(str,"hello") } : ${fn:contains(str,"hello") } <br/>
\${fn:containsIgnoreCase(str,"hello") }:
          ${fn:containsIgnoreCase(str,"hello") } <br/>
\${fn:endsWith(str,"World!") } : ${fn:endsWith(str,"World!") } <br/>
\${fn:escapeXml(str) }: ${fn:escapeXml(str) } <br/>
\${fn:indexOf(str,"o") }: ${fn:indexOf(str,"o") } <br/>
\${fn:join(arr,"_") }: ${fn:join(arr,"_") }<br/>
\${fn:length(arr) }: ${fn:length(arr) }<br/>
\${fn:replace(str,"l","L") }: ${fn:replace(str,"l","L") }<br/>
\${fn:split(str," ") }: ${fn:split(str," ") }<br/>
\${fn:startsWith(str,"Hello") } : ${fn:startsWith(str,"Hello") } <br/>
\${fn:substring(str,6,100) }: ${fn:substring(str,6,100) }<br/>
\${fn:substringAfter(str,"Hello")}: ${fn:substringAfter(str,"Hello")}<br/>
\${fn:substringBefore(str,"World") }:
       ${fn:substringBefore(str,"World") }<br/>
\${fn:toLowerCase(str) }: ${fn:toLowerCase(str) }<br/>
\${fn:toUpperCase(str) }: ${fn:toUpperCase(str) }<br/>
\${fn:trim(str) }: ${fn:trim(str) }
    </body>
</html>
```

运行的结果如图 9-11 所示。

图 9-11　运行结果

## 9.7 小　　结

本章首先对 JSTL 进行了简要介绍，讲解了 JSTL 的组成和使用方法；接下来讲解了 JSTL 的核心标签库、国际化标签库、数据库标签库、XML 标签库和函数标签库中的常用标签。在讲解的过程中结合具体的示例演示了如何利用 JSTL 进行开发。

## 9.8 习　　题

1. 什么是 JSTL？它有什么作用？
2. JSTL 中包含有哪些主要的标签库？
3. 上机实现本章所有示例。

# 第 10 章 AJAX

本章主要讲解 AJAX 技术，掌握 AJAX 开发框架，AJAX 技术在 Java Web 开发中的应用。

## 10.1 AJAX 简介

AJAX 即 Asynchronous JavaScript and XML（异步 JavaScript 和 XML），AJAX 并非缩写词，而是由 Jesse James Gaiiett 创造的名词，是指一种创建交互式网页应用的网页开发技术。

这个术语源自描述从基于 Web 的应用到基于数据的应用的转换。在基于数据的应用中，用户需求的数据如联系人列表，可以从独立于实际网页的服务端取得，并且可以被动态地写入网页中，给缓慢的 Web 应用体验着色使之像桌面应用一样。

AJAX 的核心是 JavaScript 对象 XmlHttpRequest。该对象在 Internet Explorer 5 中首次引入，它是一种支持异步请求的技术。简而言之，XmlHttpRequest 使用户可以使用 JavaScript 向服务器提出请求并处理响应，而不受到阻塞。

### 10.1.1 AJAX 包含的技术

AJAX 并非一种新的技术，而是几种原有技术的结合体。它由下列技术组合而成。
- 使用 CSS 和 XHTML 来表示。
- 使用 DOM 模型来交互和动态显示。
- 使用 XMLHttpRequest 来和服务器进行异步通信。
- 使用 JavaScript 来绑定和调用。

在上面几种技术中，除了 XmlHttpRequest 对象以外，其他的技术都是基于 Web 标准的并且已经得到了广泛使用的。XMLHttpRequest 虽然目前还没有被 W3C 所采纳，但是它已经是一个事实的标准，因为目前几乎所有的主流浏览器都支持它。

AJAX 的概念中最重要而最易被忽视的是它也是一种 JavaScript 编程语言。AJAX 的原理图如图 10-1 所示。JavaScript 是一种黏合剂使 AJAX 应用的各部分集成在一起。JavaScript 通常被服务端开发人员认为是一种企业级应用不需要使用的东西应该尽力避免。这种观点来自以前编写 JavaScript 代码的经历：繁杂而又易出错的语言。类似地，它也被认为将应用逻辑任意地散布在服务端和客户端中，这使得问题很难被发现，而且代码很难重用。在 AJAX 中，JavaScript 主要被用来传递用户界面上的数据到服务端并返回结果；XMLHttpRequest 对象用来响应通过 HTTP 传递的数据，一旦数据返回到客户端就可以立刻使用 DOM 将数据放到网页上。

XMLHttpRequest 对象在大部分浏览器上已经实现，而且拥有一个简单的接口，允许数据从客户端传递到服务端，但并不会打断用户当前的操作。使用 XMLHttpRequest 传送的数据可以是任何格式，虽然从名称上建议是 XML 格式的数据。XMLHttpRequest 对象的方法及属性分别如表 10-1、表 10-2 所示。

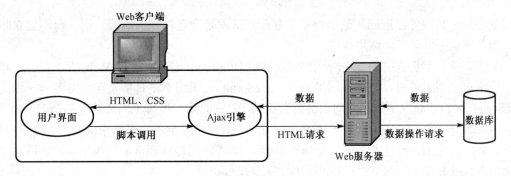

图 10-1　AJAX 原理图

表 10-1　XMLHttpRequest 对象的方法

| 方　　法 | 描　　述 |
|---|---|
| abort() | 停止当前请求 |
| getAllResponseHeaders() | 作为字符串返回完整的 headers |
| getResponseHeader("headerLabel") | 作为字符串返回单个的 header 标签 |
| open("method","URL"[,asyncFlag[,"userName"[, "password"]]]) | 设置未决的请求的目标 URL、方法和其他参数 |
| send(content) | 发送请求 |
| setRequestHeader("label", "value") | 设置 header 并和请求一起发送 |

表 10-2　XMLHttpRequest 对象的属性

| 属　　性 | 描　　述 |
|---|---|
| Onreadystatechange | 状态改变的事件触发器 |
| readyState | 对象状态(integer)：<br>0 = 未初始化<br>1 = 读取中<br>2 = 已读取<br>3 = 交互中<br>4 = 完成 |
| responseText | 服务器进程返回数据的文本版本 |
| responseXML | 服务器进程返回数据的兼容 DOM 的 XML 文档对象 |
| status | 服务器返回的状态码，如：404 = "文件未找到"、200 = "成功" |
| statusText | 服务器返回的状态文本信息 |

开发人员应该已经熟悉了许多其他 XML 相关的技术。XPath 可以访问 XML 文档中的数据，但理解 XML DOM 是必需的。类似地，XSLT 是最简单而快速地从 XML 数据生成 HTML 或 XML 的方式。许多开发人员已经熟悉 Xpath 和 XSLT，因此 AJAX 选择 XML 作为数据交换格式是有意义的。XSLT 可以被用在客户端和服务端，它能够减少大量的用 JavaScript 编写的应用逻辑。

### 10.1.2　AJAX 的运行原理

在一般的 Web 应用程序中，用户填写表单字段并单击 Submit 按钮，然后整个表单发送到服务器，服务器将它转发给处理表单的脚本（通常是 PHP 或 Java，也可能是 CGI 进程或者类似的东西）。脚本执行完成后再发送回全新的页面，该页面可能是带有已经填充某些数据的新表单的 HTML，也可能是确认页面，或者是具有根据原来表单中输入数据选择的某些选项的页面。当然，在服务器上的脚本或程序处理和返回新表单时用户必须等待，屏幕变成一片空白，等到服务器返回数据后再重新显示。这就是交互性差的原因，用户得不到立即反馈，因此感觉不同于桌面应用程序。

AJAX 基本上就是把 JavaScript 技术和 XMLHttpRequest 对象放在 Web 表单和服务器之间。当用户填写表单时，数据发送给一些 JavaScript 代码而不是直接发送给服务器。相反，JavaScript 代码捕获表单数据并向服务器发送请求。同时用户屏幕上的表单也不会闪烁、消失或延迟。换句话说，JavaScript 代码在幕后发送请求，用户甚至不知道请求的发出。更好的是，请求是异步发送的，就是说 JavaScript 代码（和用户）不用等待服务器的响应。因此用户可以继续输入数据、滚动屏幕和使用应用程序。

然后，服务器将数据返回 JavaScript 代码（仍然在 Web 表单中），后者决定如何处理这些数据。它可以迅速更新表单数据，让人感觉应用程序是立即完成的，表单没有提交或刷新而用户得到了新数据。JavaScript 代码甚至可以对收到的数据执行某种计算，再发送另一个请求，完全不需要用户干预! 这就是 XMLHttpRequest 的强大之处。它可以根据需要自行与服务器进行交互，用户甚至可以完全不知道幕后发生的一切。结果就是类似于桌面应用程序的动态、快速响应、高交互性的体验，但是背后又拥有互联网的全部强大力量。

AJAX 的原理简单来说就是：通过 XmlHttpRequest 对象来向服务器发异步请求，从服务器获得数据，然后用 JavaScript 来操作 DOM 而更新页面。这其中最关键的一步就是从服务器获得请求数据。要清楚这个过程和原理，我们必须对 XMLHttpRequest 有所了解。

XMLHttpRequest 是 AJAX 的核心机制，它是在 IE5 中首先引入的，是一种支持异步请求的技术。简单地说，也就是 JavaScript 可以及时向服务器提出请求和处理响应，而不阻塞用户，达到无刷新的效果。

所以我们先从 XMLHttpRequest 讲起，来看看它的工作原理。

首先来看 XMLHttpRequest 这个对象的属性。

它的属性如下：
- onreadystatechange 每次状态改变所触发事件的事件处理程序。
- responseText 从服务器进程返回数据的字符串形式。
- responseXML 从服务器进程返回的 DOM 兼容的文档数据对象。
- status 从服务器返回的数字代码，比如常见的 404（未找到）和 200（已就绪）。
- statusText 伴随状态码的字符串信息。
- readyState 对象状态值如下。
    - 0（未初始化）对象已建立，但是尚未初始化（尚未调用 open 方法）。
    - （初始化）对象已建立，尚未调用 send 方法。
    - （发送数据）send 方法已调用，但是当前的状态及 http 头未知。
    - （数据传送中）已接收部分数据，因为响应及 http 头不全，这时通过 responseBody 和 responseText 获取部分数据会出现错误。
    - （完成）数据接收完毕，此时可以通过 responseXml 和 responseText 获取完整的回应数据。

但是，由于各浏览器之间存在差异，所以创建一个 XMLHttpRequest 对象可能需要不同的方法。这个差异主要体现在 IE 和其他浏览器之间。下面是一个比较标准的创建 XMLHttpRequest 对象的方法。

```
function CreateXmlHttp()
{
   //非 IE 浏览器创建 XmlHttpRequest 对象
   if(window.XmlHttpRequest)
   {
    xmlhttp=new XmlHttpRequest();
   }
   //IE 浏览器创建 XmlHttpRequest 对象
   if(window.ActiveXObject)
   {
   try
   {
    xmlhttp=new ActiveXObject("Microsoft.XMLHTTP");
   }
   catch(e)
   {
   try{
    xmlhttp=new ActiveXObject("msxml2.XMLHTTP");
    }
    catch(ex){}
   }
   }
}
function Ustbwuyi()
{
   var data=document.getElementById("username").value;
   CreateXmlHttp();
   if(!xmlhttp)
   {
       alert("创建 xmlhttp 对象异常！");
       return false;
   }
   xmlhttp.open("POST",url,false);
   xmlhttp.onreadystatechange=function()
   {
       if(xmlhttp.readyState==4)
       {
           document.getElementById("user1").innerHTML="数据正在加载...";
           if(xmlhttp.status==200)
           {
               document.write(xmlhttp.responseText);
           }
       }
   }
   xmlhttp.send();
}
```

如上所示，函数首先检查 XMLHttpRequest 的整体状态并且保证它已经完成（readyStatus=4），即数据已经发送完毕。然后根据服务器的设定询问请求状态，如果一切已经就绪（status=200），那么就需要执行下面的操作。

XmlHttpRequest 有两个方法 open 和 send。

其中 open 方法指定了：

- 向服务器提交数据的类型，即 post 还是 get。
- 请求的 url 地址和传递的参数。
- 传输方式，false 为同步，true 为异步。默认为 true。如果是异步通信方式，客户机就不等待服务器的响应；如果是同步方式，客户机就要等到服务器返回消息后才去执行其他操作。根据实际需要来指定同步方式，在某些页面中，可能会发出多个请求，甚至是有组织、有计划、有队形、大规模、高强度的 request，而后一个会覆盖前一个，这个时候当然要指定同步方式。

Send 方法用来发送请求。

知道了 XMLHttpRequest 的工作流程，我们可以看出，XMLHttpRequest 是完全用来向服务器发出一个请求的，它的作用也局限于此，但它的作用是整个 AJAX 实现的关键，因为 AJAX 无非是两个过程，发出请求和响应请求。并且它完全是一种客户端的技术。XMLHttpRequest 正是处理了服务器端和客户端通信的问题所以才会如此的重要。

现在，我们对 AJAX 的原理大概可以有一个了解了。我们可以把服务器端看成一个数据接口，它返回的是一个纯文本流，当然，这个文本流可以是 XML 格式，可以是 HTML，可以是 JavaScript 代码，也可以只是一个字符串。这时候，XMLHttpRequest 向服务器端请求这个页面，服务器端将文本的结果写入页面，这和普通的 Web 开发流程是一样的，不同的是，客户端在异步获取这个结果后，不是直接显示在页面，而是先由 JavaScript 来处理，然后再显示在页面。至于现在流行的很多 AJAX 控件，比如 Magicajax 等，可以返回 DataSet 等其他数据类型，这只是将这个过程封装了的结果，本质上它们并没有什么太大的区别。

## 10.2 AJAX 开发框架

AJAX 实质上也是遵循 Request/Server 模式，所以 AJAX 开发框架基本的流程也是：对象初始化；发送请求；服务器接收；服务器返回；客户端接收；修改客户端页面内容。只不过这个过程是异步的。

为了让 JavaScript 可以向服务器发送 HTTP 请求，必须使用 XMLHttpRequest 对象。使用之前，要先将 XMLHttpRequest 对象实例化。不同的浏览器对 XMLHttpRequest 的实例化过程实现不同，IE 以 ActiveX 控件的形式提供，而 Mozilla 等浏览器则直接以 XMLHttpRequest 类的形式提供。为了让编写的程序能够跨浏览器运行，要这样写：

```
if (window.XMLHttpRequest) { // Mozilla, Safari, …
    http_request = new XMLHttpRequest();
}
else if (window.ActiveXObject) { // IE
    http_request = new ActiveXObject("Microsoft.XMLHTTP");
}
```

有些版本的 Mozilla 浏览器处理服务器返回的未包含 XML mime-type 头部信息的内容时会出错。因此，要确保返回的内容包含 text/xml 信息。

```
http_request = new XMLHttpRequest();
http_request.overrideMimeType('text/xml');
```

接下来要指定当服务器返回信息时客户端的处理方式。只要将相应的处理函数名称赋给 XMLHttpRequest 对象的 onreadystatechange 属性就可以了。比如：

```
http_request.onreadystatechange = processRequest;
```

需要指出的是，这个函数名称不加括号，不指定参数。也可以用 JavaScript 即时定义函数的方式定义响应函数。比如：

```
http_request.onreadystatechange = function() { };
```

指定响应处理函数之后，就可以向服务器发出 HTTP 请求了。这一步调用 XMLHttpRequest 对象的 open 和 send 方法。

```
http_request.open('GET', 'http://www.example.org/some.file', true);
http_request.send(null);
```

open 的第一个参数是 HTTP 请求的方法，为 Get、Post 或者 Head。

open 的第二个参数是目标 URL。基于安全考虑，这个 URL 只能是同网域的，否则会提示"没有权限"的错误。这个 URL 可以是任何的 URL，包括需要服务器解释执行的页面，不仅仅是静态页面。目标 URL 处理请求 XMLHttpRequest 请求则跟处理普通的 HTTP 请求一样，比如 JSP 可以用 request.getParameter("")或者 request.getAttribute("")来取得 URL 参数值。

open 的第三个参数只是指定在等待服务器返回信息的时间内是否继续执行下面的代码。如果为 True，则不会继续执行，直到服务器返回信息。默认为 True。

按照顺序，open 调用完毕之后要调用 send 方法。send 的参数如果是以 Post 方式发出的话，可以是任何想传给服务器的内容。不过，跟 form 一样，如果要传文件或者 Post 内容给服务器，必须先调用 setRequestHeader 方法，修改 MIME 类别。如下：

```
http_request.setRequestHeader("Content-Type","
        application/x-www-form-urlencoded");
```

这时资料则以查询字符串的形式列出，作为 sned 的参数，例如：

```
name=value&anothername=othervalue&so=on
```

处理服务器返回的信息，在第二步我们已经指定了响应处理函数，这一步，来看看这个响应处理函数都应该做什么。

首先，它要检查 XMLHttpRequest 对象的 readyState 值，判断请求目前的状态。参照前文的属性表可以知道，readyState 值为 4 的时候，代表服务器已经传回所有的信息，可以开始处理信息并更新页面内容了。如下：

```
if (http_request.readyState == 4) {
    // 信息已经返回，可以开始处理
} else {
    // 信息还没有返回，等待
}
```

服务器返回信息后，还需要判断返回的 HTTP 状态码，确定返回的页面没有错误。所有的状态码都可以在 W3C 的官方网站上查到。其中，200 代表页面正常。

```
        if (http_request.status == 200) {
            // 页面正常,可以开始处理信息
        } else {
            // 页面有问题
        }
```

XMLHttpRequest 对成功返回的信息有以下两种处理方式。

responseText:将传回的信息当字符串使用。

responseXML:将传回的信息当 XML 文档使用,可以用 DOM 处理。

总结上面的步骤,我们整理出一个初步的可用的开发框架,供以后调用。这里,将服务器返回的信息用 window.alert 以字符串的形式显示出来。

```
<script language="javascript">
var http_request = false;
function send_request(url) {//初始化、指定处理函数、发送请求的函数
http_request = false;
//开始初始化 XMLHttpRequest 对象
if(window.XMLHttpRequest) { //Mozilla 浏览器
http_request = new XMLHttpRequest();
if (http_request.overrideMimeType) {//设置 MIME 类别
http_request.overrideMimeType("text/xml");
}
}
else if (window.ActiveXObject) { // IE 浏览器
    try {
        http_request = new ActiveXObject("Msxml2.XMLHTTP");
    } catch (e) {
        try {
            http_request = new ActiveXObject("Microsoft.XMLHTTP");
        } catch (e) {}
    }
}
if (!http_request) { // 异常,创建对象实例失败
    window.alert("不能创建 XMLHttpRequest 对象实例.");
    return false;
}
http_request.onreadystatechange = processRequest;
// 确定发送请求的方式和 URL 以及是否同步执行下段代码
http_request.open("GET", url, true);
http_request.send(null);
}
// 处理返回信息的函数
function processRequest() {
if (http_request.readyState == 4) { // 判断对象状态
if (http_request.status == 200) { // 信息已经成功返回,开始处理信息
alert(http_request.responseText);
} else { //页面不正常
    alert("您所请求的页面有异常。");
}
```

```
        }
    }
</script>
```

## 10.3　AJAX 应用

我们在做验证码的时候往往要反作弊，验证有时故意加入多的干扰因素，这时验证码显示得不很清楚，用户经常输入错误。这样不但要重新刷新页面，导致用户因没有看清楚验证码而重填，而不是修改，而且如果没有用 session 保存下用户输入的其他数据的话（如姓名），用户刚刚输入的内容也不存在了，这样给用户造成不好的体验。

本例在原有验证方式基础之上增加一段 JS，通过 XMLHTTP 来获取返回值，以此来验证是否有效，这样即使用户浏览器不支持 JS，也不会影响他的正常使用了。

为了防止作弊，当用户连接 3 次输入错误时则重载一下图片，这样也避免了用户因为图片上的验证码辨认不清而始终无法输入正确。

本例还特别适合检验用户名是有效，只要从后台做个 SQL 查询，返回一个值或是 XML 即可。

本例的优点在于非常方便用户输入，而且减少对服务器端的请求，可以说既改善了用户体验又会节省带宽成本，但相应地要在页面上增加一段 JavaScript 代码。在目前网速越来越快、人们要求便捷舒适的今天，似乎我们更应注意提供给用户良好的使用感受。

### 10.3.1　基于 JSP 的 AJAX

这个例子是关于输入校验的问题，在用户注册时，表单需要到数据库中对该用户登录名进行唯一性确认，然后才能继续往下注册。这种校验需要访问后台数据库，但又不希望用户在这里提交后等待，这种场景适用 AJAX 技术实现。

在客户端显示 UI 的 JSP 如下：

```
<%@ page language="java" import="java.util.*" pageEncoding="utf-8"%>
<!DOCTYPE HTML PUBLIC "-//W3C//DTD HTML 4.01 Transitional//EN">
<html>
  <head>
    <title>My JSP 'zhuce.jsp' starting page</title>
    <meta http-equiv="pragma" content="no-cache">
    <meta http-equiv="cache-control" content="no-cache">
    <meta http-equiv="expires" content="0">
    <meta http-equiv="keywords" content="keyword1,keyword2,keyword3">
    <meta http-equiv="description" content="This is my page">
  </head>
<script type="text/javascript">
    var http_request = false;
    function send_request(url) {//初始化、指定处理函数、发送请求的函数
        http_request = false;
        //开始初始化 XMLHttpRequest 对象
        if (window.XMLHttpRequest) { //Mozilla 浏览器
            http_request = new XMLHttpRequest();
```

```
            if (http_request.overrideMimeType) {//设置MIME类别
                http_request.overrideMimeType('text/xml');
            }
        } else if (window.ActiveXObject) { // IE 浏览器
            try {
                http_request = new ActiveXObject("Msxml2.XMLHTTP");
            } catch (e) {
                try {
                    http_request = new ActiveXObject("Microsoft.XMLHTTP");
                } catch (e) {
                }
            }
        }
        if (!http_request) { // 异常,创建对象实例失败
            window.alert("不能创建 XMLHttpRequest 对象实例.");
            return false;
        }
        http_request.onreadystatechange = processRequest;
        // 确定发送请求的方式和 URL 以及是否同步执行下段代码
        http_request.open("GET", url, true);
        http_request.send(null);
    }
    // 处理返回信息的函数
    function processRequest() {
        if (http_request.readyState == 4) { // 判断对象状态
            if (http_request.status == 200) { // 信息已经成功返回,开始处理信息
                alert(http_request.responseText);
            } else { //页面不正常
                alert("您所请求的页面有异常。");
            }
        }
    }
    function userCheck() {
        var f = document.form1;
        var username = f.username.value;
        if (username == "") {
            window.alert("The user name can not be null!");
            f.username.focus();
            return false;
        } else {
            send_request('checkName.jsp?username=' + username);
        }
    }
</script>
<body>
    <form name="form1" action="" method="post">
        User Name:<input type="text" name="username" value="">  <input
            type="button" name="check" value="check" onClick="userCheck()">
```

```
            <input type="submit" name="submit" value="submit">
    </form>
</body>
</html>
<%@ page language="java" import="java.util.*,java.sql.*,
            java.text.*" pageEncoding="UTF-8"%>
<!DOCTYPE HTML PUBLIC "-//W3C//DTD HTML 4.01 Transitional//EN">
<html>
  <head>
    <title>会员注册</title>
    <meta http-equiv="pragma" content="no-cache">
    <meta http-equiv="cache-control" content="no-cache">
    <meta http-equiv="expires" content="0">
    <meta http-equiv="keywords" content=
            "keyword1,keyword2,keyword3">
    <meta http-equiv="description" content="This is my page">
    <link rel="stylesheet" type="text/css" href="bbs03.css">
</head>

<body>

<script Language="JavaScript">
<!--
function getResult() {
    var url = "checkname.jsp?name="+ encodeURI(encodeURI
            (document.registryForm.username.value));
    if (window.XMLHttpRequest) {
        req = new XMLHttpRequest();
    } else if (window.ActiveXObject) {
        req = new ActiveXObject("Microsoft.XMLHTTP");
    }
    if (req) {
        req.onreadystatechange = complete;
        req.open("get", url, true);
        req.setRequestHeader("Content-Type",
                "application/x-www-form-urlencoded");
        req.send(null);
    }
}
function complete() {
    if (req.readyState == 4) {
        if (req.status == 200) {
            document.getElementById("checkname").innerHTML =
                    req.responseText;
        }
    }
}
```

```javascript
function isspacestring(mystring)
{ var istring=mystring;
  var temp,i,strlen;
  temp=true;
  strlen=istring.length;
  for (i=0;i<strlen;i++)
  {
    if ((istring.substring(i,i+1)!=" ")&(temp))
      { temp=false; }
  }
  return temp;
}

function firstisspace(mystring)
{ var istring=mystring;
  var temp,i,strlen;
  temp=false;
    if (istring.substring(0,1)==" ")
      { temp=true; }
  return temp;
}

function isemail(mystring){
  var istring=mystring;
  var atpos=mystring.indexOf("@");
  var temp=true;
  if (atpos==-1) //email中没有@符号：不正确的EMAIL
  {
    temp=false;
  }
  return temp;
}

function checkForm(){

  if ((registryForm.username.value == "")|
          (firstisspace(registryForm.username.value))){
    alert("请输入用户名.不能以空格开头.");
    registryForm.username.focus();
    return false;
  }

  if (registryForm.username.value.length > 20){
    alert("用户名长度应小于20个字符或数字.");
    registryForm.username.focus();
    return false;
  }
```

```
        if ((registryForm.userpassword.value == "")|
                (isspacestring(registryForm.userpassword.value))){
            alert("请输入密码.");
            registryForm.userpassword.focus();
            return false;
        }

        if (registryForm.userpassword.value.length > 10){
            alert("密码长度应小于10.");
            registryForm.user_password.focus();
            return false;
        }

        if (registryForm.userpassword.value !=
                registryForm.userpassword2.value){
            alert("两次密码不相同,请重新输入密码.");
            registryForm.userpassword.focus();
            return false;
        }
        return true;
    }
    //-->
</script>

<div align="center">
<table bgcolor="#3299CC" border="1" width="600" bordercolorlight=
    "#000000" cellspacing="0" cellpadding="0" bordercolordark="#FFFFFF">
    <tr>
        <td width="600" align="center">
            <form method="post" action="registerServlet" onSubmit=
                "return checkForm()" name="registryForm">
                <p align="center" style="margin-top: 0; margin-bottom:
                    0"><b><font size="4"><br>
                    BBS会员注册   </font></b><font color="#FF0000">
                    (*</font>不能为空<font color="#FF0000">)</font></p>
                <div align="center">
                    <center>
                    <table bgcolor="#E6E8FA" border="0" width="95%" height="85">
                        <tr>
                            <td width="540" height="41">
                                <p style="margin-top: 0; margin-bottom: 0">用户名:
                                    <input class="intext" type="text" name="username"
                                        size="20" > 
                                    <font color="#FF0000">*</font><span id="checkname">
                                        <input type="button" value="唯一性验证"
                                    onclick="getResult()"></span></p>
                                <p>密   码:
                                    <input class="intext" type="password" name=
                                        "userpassword" size="10">
```

```html
          <font color="#FF0000">*</font>密码长度应小于10!!</p>
        <p> 重复密码:
          <input class="intext" type="password" name=
                  "userpassword2" size="10">
          <font color="#FF0000">*</font></p>
        <p style="margin-top: 0">E_MAIL  
          <input class="intext" type="text" name="useremail"
                  size="30"></p>
        <p style="margin-top: 0">HOMEPAGE:
          <input class="intext" type="text" name="userhomepage"
                  size="50" value="http://"></p>
        <p style="margin-top: 0">主页名称:
          <input class="intext" type="text" name="hpname"
                  size="20"></p>
        <p style="margin-top: 0">生    日:
          <input class="intext" type="text" name="userbirthday"
                  size="11" value="99-05-26">  格式:
                  yy-mm-dd </p>
        <p>    <br>
          性   别:
          <input type="radio" value="男" name="usersex" checked>
          男
          <input type="radio" value="女" name="usersex" checked>
          女                   </p>
        <p>来   自:
          <input class="intext" type="text" name="comefrom"
                  size="20"></p>
        <p style="margin-top: 0; margin-bottom: 0">签名:</p>
        <p style="margin-top: 0; margin-bottom: 0">
          <textarea class="intext" rows="4" name="usersign"
                  cols="36"></textarea>
          <br>
          <input class="buttonface" type="submit" value="提交"
                  name="B1">  
          <input class="buttonface" type="reset" value="重填"
                  name="B2"> 
        </p>
      </td>
    </tr>
   </table>
   </center>
   </div>
   </form>
   </td>
  </tr>
 </table>
</div>
</body>
</html>
```

运行结果如图 10-2 所示。

图 10-2　客户端显示

服务器端处理逻辑的 JSP 如下：

```jsp
<%@ page contentType="text/html; charset=utf-8" language="java" errorPage="" %>
<%
    String username=request.getParameter("name");
    //String username = URLDecoder.decode(str,"utf-8");
    boolean f = userdao.getUserInfo(username);
    String rs = "";
    if(f){
        rs = "该名字已使用，请重新输入！";
    }else{
        rs = "该名字可以使用！";
    }
    out.print(rs);
%>
```

运行结果如图 10-3 所示。

图 10-3　服务端处理逻辑

## 10.3.2　基于 Servlet 的 AJAX

BBS 案例中，主题回复，可以利用 AJAX 技术实现页面局部刷新。

客户端的 JSP 页面如下：

```jsp
<%@ page language="java" import="net.bbs.bean.*,java.util.*" pageEncoding=
    "UTF-8"%>
<!DOCTYPE HTML PUBLIC "-//W3C//DTD HTML 4.01 Transitional//EN">
<html>
<head>
<title>加帖</title>
<meta http-equiv="Content-Type" content="text/html; charset=utf-8">
<meta http-equiv="pragma" content="no-cache">
<meta http-equiv="cache-control" content="no-cache">
<meta http-equiv="expires" content="0">
<meta http-equiv="keywords" content="keyword1,keyword2,keyword3">
<meta http-equiv="description" content="This is my page">
<link rel="stylesheet" type="text/css" href="bbs03.css">
</head>
<%
BbsreTop bbsre = (BbsreTop)request.getAttribute("bbsreTop");
String boardname = bbsre.getBoardname();
```

```
        int boardid = bbsre.getBoardid();
%>
<body>
    <script Language="JavaScript">
<!--
function getResult() {
        var url = "/bbs/jsp/ajaxBbsreServlet" ;
        var pm = "boardid="+
            encodeURI(encodeURI(bbsreform.boardid.value));
        pm = pm +"&parentid="+
            encodeURI(encodeURI(bbsreform.parentid.value));
        pm = pm +"&bbsid="+ encodeURI(encodeURI(bbsreform.bbsid.value));
        pm = pm +"&expression="+
            encodeURI(encodeURI(bbsreform.expression.value));
        pm = pm +"&bbstopic="+
            encodeURI(encodeURI(bbsreform.bbstopic.value));
        pm = pm +"&bbscontent="+
            encodeURI(encodeURI(bbsreform.bbscontent.value));
        pm = pm +"&boardname="+
            encodeURI(encodeURI(bbsreform.boardname.value));
        if (window.XMLHttpRequest) {
            req = new XMLHttpRequest();
        } else if (window.ActiveXObject) {
            req = new ActiveXObject("Microsoft.XMLHTTP");
        }
        if (req) {
            req.onreadystatechange = complete;
            req.open("post", url, true);
            req.setRequestHeader("Content-Type","application/
                x-www-form-urlencoded");
            req.send(pm);
        }
    }
    function complete() {
        if (req.readyState == 4) {
            if (req.status == 200) {
                document.getElementById("bbs").innerHTML = req.responseText;
            }
        }
    }

function clear(){
    document.bbsreform.bbstopic.value="";
    document.bbsreform.bbscontent.value="";
}

function check(){
    if(document.bbsreform.bbstopic.value == ""){
        alert("主题不能为空!");
        bbsreform.bbstopic.focus();
```

```
            return false;
        }
        else if(document.bbsreform.bbscontent.value == ""){
            alert("内容不能为空!");
            bbsreform.bbscontent.focus();
            return false;
        }
    }
//-->
</script>
    <!-------------主帖----------------->
    <table border="0" width="100%" cellspacing="0" cellpadding="0">
        <tr>
            <td width="100%">
                <p align="center">
                    <b>主题:<%=bbsre.getBbstopic()%></b>
                </p>
                <hr color="#000080" size="1">
                <p><%=bbsre.getUsername()%>
                    于<%=bbsre.getDateandtime()%>发表于:<b><%=boardname%></b>
                </p>
                <p>
                    内容:<%=bbsre.getBbscontent()%></p>
                <p align="left">
                    <font color="#000080">用户签名:<%=bbsre.getUsersign()%></font>
                </p>
                <hr color="#000080"></td>
        </tr>
    </table>
    <!-------------所有跟帖----------------->
    <%
        List list = bbsre.getList();
        for (int i = 0;i<list.size();i++) {
            Bbs bbs = (Bbs)list.get(i);
    %>

    <table border="0" width="100%" cellspacing="0" cellpadding="0">
        <tr>
            <td width="100%">
                <p><%=bbs.getUsername()%>
                    于<%=bbs.getDateandtime()%>
                    发表于:<b><%=boardname%></b>
                </p>
                <p>
                    <b>主题:<%=bbs.getBbstopic()%></b>
                </p>
                <p>
                    内容:<%=bbs.getBbscontent()%></p>
                <p align="left">
```

```
                    <font color="#000080">用户签名: <%=bbs.getUsersign()%></font>
                    </p>
                    <hr color="#000080" size="1"></td>
        </tr>
</table>
    <%}%>



    <!--------------AJAX 跟帖----------------->
    <div id="bbs"></div>
    <!--------------AJAX 跟帖----------------->

<div align="center">
    <center>
        <table border="1" width="500" bordercolorlight="#000000"
            cellspacing="0" cellpadding="0"
                bordercolordark="#FFFFFF">
            <tr>
                <td width="100%" bgcolor="#99CC66">
                    <p align="center">
                        <font color="#FFFFFF">我 要 回 复</font>
                </td>
            </tr>
            <tr>
            <td width="100%" bgcolor="#FFFFFF" valign="top">
            <div align="center">
            <center>
            <table border="0" width="440">
            <tr>
            <td width="100%" valign="top">
            <!-- postBbsreServlet -->
            <form method="post" action="/bbs/jsp/ajaxBbsreServlet"
                name="bbsreform" target="_self">
            <p style="margin-top: 0; margin-bottom: 0">
            <b><font color="#008000">
            版面:</font>
            </b> <a href="preBbslistServlet? boardid=<%=boardid%>">
            <%=boardname%></a>
            </p>

            <p align="left" style="margin-top: 0; margin-bottom:0">
            <b><font color="#008000">主题:
            </font>
            </b> <input type="text" name="bbstopic" size="50">
            </p>
            <p align="left" style="margin-top: 0; margin-bottom: 0">
            <font color="#008000"><b>
            表情:</b>
            </font> <input type="radio" value="001.gif" name=
                "expression"checked>
            <img border="0" src="images/001.gif">
```

```html
<input type="radio" value="002.gif" name="expression">
<img border="0" src="images/002.gif">
<input type="radio" value="003.gif" name="expression">
<img border="0" src="images/003.gif" width="20"height="20">
<input type="radio" value="004.gif" name="expression">
<img border="0" src="images/004.gif">
<input type="radio" value="005.gif" name="expression">
<img border="0" src="images/005.gif">
<inputtype="radio" value="006.gif" name="expression">
<img border="0" src="images/006.gif">
<inputtype="radio" value="007.gif" name="expression">
<img border="0" src="images/007.gif">
<input type="radio" value="008.gif" name="expression">
<img border="0" src="images/008.gif">
<input type="radio" value="009.gif" name="expression">
<img border="0" src="images/009.gif">
</p>
<p align="left" style="margin-top:0; margin-bottom: 0">
         
<input type="radio" value="010.gif" name="expression">
<img border="0" src="images/010.gif">
<input type="radio" value="011.gif" name="expression">
<img border="0" src="images/011.gif">
<input type="radio" value="012.gif" name="expression">
<img border="0" src="images/012.gif">
<input type="radio" value="013.gif" name="expression">
<img border="0" src="images/013.gif">
<input type="radio" value="014.gif" name="expression">
<img border="0" src="images/014.gif">
</p>
<p align="left" style="margin-top: 0; margin-bottom: 0">
<b><font color="#008000">
内容:</font>
</b><b><font color="#008000">
<p align="left" style="margin-top:0; margin-bottom:0">
<textarea rows="4" name="bbscontent" cols="50">
</textarea>
</p> </font>
</b> <input class="buttonface" type="button" value="发表"
    name="B1" onclick="getResult()">
     
<input class="buttonface" type="reset" value="取消"
    name="B2">
</p>

<input type="hidden" name="bbsid"value="<%=bbsre.getBbsid()%>">
<input type="hidden" name="boardid" value= "<%=bbsre.get-
    Boardid()%>">
<input type="hidden" name="parentid" value="<%=bbsre.
    getParentid()%>">
```

```html
                    <input type="hidden" ame="boardname" value="<%=
                        boardname%>">
                </form></td>
            </tr>
        </table>
        </center>
        </div></td>
        </tr>
    </table>
    </center>
    </div>
</body>
</html>
```

服务器端业务逻辑的 Servlet 如下：

```java
public class AjaxBbsreServlet extends HttpServlet {
    protected void doGet(HttpServletRequest req, HttpServletResponse resp)
            throws ServletException, IOException {
        doPost(req, resp);
    }

    @Override
    protected void doPost(HttpServletRequest request,
            HttpServletResponse response)
            throws ServletException, IOException {
        request.setCharacterEncoding("utf-8");
        response.setCharacterEncoding("utf-8");

        PrintWriter out = response.getWriter();
        HttpSession session = request.getSession();
        Integer userid = (Integer)session.getAttribute("userid");

        String userip = request.getRemoteAddr();

        String whereTo = "";

        if (userid != null) {
            String str = request.getParameter("state");

            String str1=request.getParameter("boardid");
            String boardidSTR = URLDecoder.decode(str1,"utf-8");
            int boardid=java.lang.Integer.parseInt(boardidSTR);

            String str2=request.getParameter("parentid");
            String parentidSTR = URLDecoder.decode(str2,"utf-8");
            int parentid=java.lang.Integer.parseInt(parentidSTR);

            String str3=request.getParameter("bbsid");
```

```java
        String bbsidSTR = URLDecoder.decode(str3,"utf-8");
        int bbsid=java.lang.Integer.parseInt(bbsidSTR);

        //开始加帖操作
        int child=0;
        String str4=request.getParameter("expression");
        String  expression= URLDecoder.decode(str4,"utf-8");

        String str5=request.getParameter("bbstopic");
        String bbstopic = URLDecoder.decode(str5,"utf-8");
        System.out.println("ajaxservlet...presave...
                bbstopic:"+bbstopic);

        String str6=request.getParameter("bbscontent");
        String bbscontent = URLDecoder.decode(str6,"utf-8");
        System.out.println("ajaxservlet...presave...
                bbscontent:"+bbscontent);

        String str7=request.getParameter("boardname");
        String boardname = URLDecoder.decode(str7,"utf-8");

        int bbshits=0;
        int bbslength=bbscontent.length();

        BbsDAO bbsdao = new BbsDAO();
        Bbs bbs = bbsdao.addAjaxBbsre(parentid,bbsid,boardid,child,
                expression,userid,userip,bbstopic,
                bbscontent,bbshits,bbslength);

        //response.setContentType("text/xml");
        response.setHeader("Cache-Control", "no-cache");
        response.setCharacterEncoding("UTF-8");

        String rs = "<table border='0' width='100%' cellspacing='0'
            cellpadding='0'><tr><td width='100%'>";
        rs+="<p>"+bbs.getUsername()+"于"+bbs.getDateandtime()+"发表于:
            <b>"+boardname+"</b></p><p><b>主题："+bbstopic+"</b></p>
            <p>内容："+bbs.getBbscontent()+"</p>";
        rs+="<p align='left'><font color='#000080'>用户签名：
            "+bbs.getUsersign()+"</font></p><hr color='#000080'
            size='1'></td></tr></table>";
        out.write(rs);
        out.close();

    } else {
        whereTo = "/jsp/login.jsp";
```

```
                request.getRequestDispatcher(whereTo).forward(request, response);
            }
        }
    }
```

这个类也十分简单,首先是从 request 里取得 state 参数,然后根据 state 参数生成相应的 XML 文件,最后将 XML 文件输出到 PrintWriter 对象里。

```xml
Web.xml
<?xml version="1.0" encoding="UTF-8"?>
<web-app version="3.0"
    xmlns="http://java.sun.com/xml/ns/javaee"
    xmlns:xsi="http://www.w3.org/2001/XMLSchema-instance"
    xsi:schemaLocation="http://java.sun.com/xml/ns/javaee
    http://java.sun.com/xml/ns/javaee/web-app_3_0.xsd">

    <servlet>
        <servlet-name>ajaxBbsreServlet</servlet-name>
        <servlet-class>net.bbs.servlet.AjaxBbsreServlet
            </servlet-class>
    </servlet>
    <servlet-mapping>
        <servlet-name>ajaxBbsreServlet</servlet-name>
        <url-pattern>/servlet/ajaxBbsreServlet </url-pattern>
    </servlet-mapping>
     <welcome-file-list>
        <welcome-file>index.jsp</welcome-file>
     </welcome-file-list>
</web-app>
```

AJAX 的初始页面及应用结果如图 10-4、图 10-5 所示。

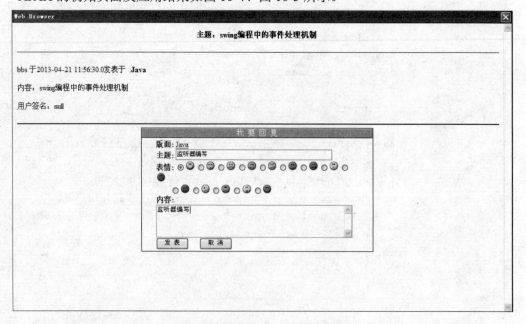

图 10-4　AJAX 的初始页面

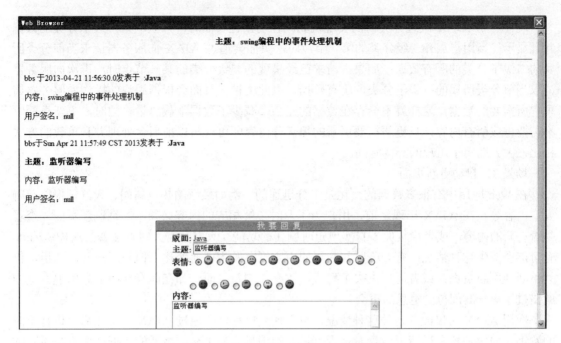

图 10-5 基于 Servlet 的 AJAX 应用结果

### 10.3.3 AJAX 的应用场景

AJAX 不是万能的，在适合的场合使用 AJAX，才能充分发挥它的长处，改善系统性能和用户体验，绝不可以为了技术而滥用。AJAX 的特点在于异步交互，动态更新 Web 页面，因此它的适用范围是交互较多、频繁读取数据的 Web 应用。现在来看几个 AJAX 的应用实例，读者可以了解如何使用 AJAX 技术改进现有的 Web 应用系统。

场景 1．数据验证

在填写表单内容时，需要保证数据的唯一性（例如新用户注册填写的用户名），因此必须对用户输入的内容进行数据验证。数据验证通常有两种方式：一种是直接填写，然后提交表单，这种方式需要将这个页面提交到服务器端进行验证，整个过程不仅时间长而且造成了服务器不必要的负担。第二种方式是改进了的验证过程，用户可以通过单击相应的验证按钮，打开新窗口查看验证结果，但是这样需要新开一个浏览器窗口或者对话框，还需要专门编写验证的页面，比较耗费系统资源。而使用 AJAX 技术，可以由 XMLHttpRequest 对象发出验证请求，根据返回的 HTTP 响应判断验证是否成功，整个过程不需要弹出新窗口，也不需要将整个页面提交到服务器，快速而又不加重服务器负担。

场景 2．按需取数据

分类树或者树形结构在 Web 应用系统中使用得非常广泛，例如部门结构、文档分类结构等常常使用树形呈现。以前每次对分类树的操作都会引起页面重载，为了避免这种情况出现，一般不采用每次调用后台的方式，而是将分类结果中的数据一次性读取出来并写入数组，然后根据用户的操作，用 JavaScript 来控制节点的呈现，这样虽然解决了操作响应速度，不重复载入页面以及避免向服务器频繁发送请求的问题，但是如果用户不对分类进行操作或者只对分类树中的一部分数据进行操作的话（这种情况很普遍的），那么读取的数据中就会有相当大的冗余，浪费了用户的资源。特别是在分类结构复杂、数据庞大的情况下，这种弊端就更加明显了。

现在应用 AJAX 改进分类树的实现机制。在初始化页面时，只获取第一级子分类的数据并且显示；当用户点开一级分类的第一节点时，页面会通过 AJAX 向服务器请求当前分类所属的二级子分类的所有数据；如果再请求已经呈现的二级分类的某一节点时，再次向服务器请求当前分类所属的三级子分类的所有数据，以此类推。页面会根据用户的操作向服务器请求它所需要的数据，这样就不会存在数据的冗余，减少了数据下载总量。同时，更新页面时不需要重载所有内容，只更新需要更新的那部分内容即可，与以前后台处理并且重载的方式相比，大大缩短了用户的等待时间。

场景3．自动更新页面

在 Web 应用中有很多数据的变化是十分迅速的，例如最新的热点新闻、天气预报以及聊天室内容等。在 AJAX 出现之前，用户为了及时了解相应的内容必须不断地刷新页面，查看是否有新的内容，或者由页面本身实现定时刷新的功能（大多数聊天室页面就是这样做的）。有可能会发生这种情况；有一段时间网页的内容没有发生任何变化，但是用户并不知道，仍然不断地刷新页面；或者用户失去了耐心，放弃了刷新页面，却在此有新的消息出现，这样就错过了第一时间得到消息的机会。

应用 AJAX 可以改善这种这种情况，页面加载以后，会通过 AJAX 引擎在后台进行定时的轮询，向服务器发送请求，查看是否有最新的消息。如果有则将新的数据（而不是所有数据）下载并且在页面上进行动态更新，通过一定的方式通知用户（实现这样的功能正是 JavaScript 的强项）。这样既避免了用户不断手工刷新页面的不便，也不会因为重复刷新页面造成资源浪费。

## 10.4 小　　结

本章首先对 AJAX 进行了简要介绍，讲解了 AJAX 的技术构成以及 AJAX 的运行原理；接下来讲解了 AJAX 的开发框架；最后讲解了 AJAX 应用，基于 JSP 的 AJAX 应用和基于 Servlet 的 AJAX 应用。

## 10.5 习　　题

1. 简述 AJAX 主要涉及哪些方面的技术。
2. 描述 AJAX 的运行过程的步骤。

# 第 11 章 综合案例

本章将综合各章概念、技术和方法，把这些概念、技术和方法运用到实际应用当中，使读者 Java Web 技术和应用有更深刻的理解。本章主要介绍了两个例子：博客网站和网上书店。

## 11.1 博客网站

### 11.1.1 系统功能

博客正在改变社会交流方式。目前，全球参与博客的人数已经达到千万之众，而且每几秒钟都有一名新的博客用户加入进来。一般来说，一个博客系统包含前台和后台。前台主要功能有：显示菜单栏，文章列表，文章类型列表，友情链接，显示文章；后台主要功能有两个，一个是添加菜单，添加文章、分类、友情链接、用户；一个是管理菜单，管理文章、分类、用户、链接、评论。

### 11.1.2 数据库设计

本系统采用 MySQL 数据库，在数据库中使用 article 来储存现有的博客文章信息。表的结构如表 11-1 所示。

表 11-1 博客文章表 article

| 字 段 名 称 | 数 据 类 型 | 说 明 |
| --- | --- | --- |
| Id | Int(10) | 博客编号，自动增加（主键） |
| Title | Varchar(45) | 博客的标题 |
| classid | Int(10) | 博客类别 id |
| zhengwen | Varchar(45) | 正文 |
| Blogid | Varchar（45） | 发表者 |

使用用户表 Blog 来储存用户的信息，表的结构如表 11-2 所示。

表 11-2 用户表 Blog

| 字 段 名 称 | 数 据 类 型 | 说 明 |
| --- | --- | --- |
| Id | Int(10) | 编号，自动增加（主键） |
| Username | Varchar(45) | 用户名 |
| Password | Varchar(128) | 用户密码 |
| Subject | Varchar(45) | 博客标题 |
| Visitcount | Int(10) | 访问次数 |

使用博客评论表 comment 来储存博客评论信息，表的结构如表 11-3 所示。
使用文章分类表 classes 来储存新闻类别信息，表的结构如表 11-4 所示。

表 11-3 博客评论表 comment

| 字 段 名 称 | 数 据 类 型 | 说　　明 |
|---|---|---|
| newsid | Int(10) | 新闻 id |
| ComTime | timestamp | 评论时间 |
| ComPerson | Varchar(45) | 评论者 |
| Content | Varchar(45) | 评论内容 |

表 11-4 文章分类表 classes

| 字 段 名 称 | 数 据 类 型 | 说　　明 |
|---|---|---|
| classid | Int(10) | 类别 id 自动增长 |
| classname | Varchar(45) | 类别名称 |

## 11.1.3　系统主要功能的实现

**1. 博客文章的显示**

任何用户在浏览器输入网址后进入本站界面，即可查看博客文章标题及发布时间，博客文章的评论等，单击任意一条博客文章标题即可查看博客文章内容。

显示一条博客的主要代码如下：

```jsp
<%
    request.setCharacterEncoding("gbk");
    SearchNews sh = new SearchNews();
    News n=sh.getNews(1);
    if(n != null)
    {
%>
<% out.println(n.getBiaoti()); %>
<% out.println(n.getLaiyuan()); %>
<% out.println(n.getTitle()); %></td>
<% out.println(n.getzhengwen()); %>
<% out.println(n.Blogid()); %>
<% out.println(n.getClassid()); %>
<% } %>
<%out.println(n.getBanquanxinxi()); %>
<% if(n.getComcount()!=0) { %>
<a href="commentofnew.jsp?newid=<% out.print(n.getNewsid()); %>"> <em>评论
      <% out.print(n.getComcount()); %>    条>></em> </a>
      <% } else { %>   <em>评论
      <% out.print(n.getComcount()); %>    条>></em> <% } %></td>
   </tr>
<p><hr>我要留言：</hr>
<form action="AddCommentOfNews" method="post" enctype=
        "multipart/form-data" name="form1" id="form1">
    <textarea name="commenttext" rows="15" cols="61"></textarea>
    <input name="newsid" type="hidden" value="<%out.print(n.getNewsid()); %>"/>
    <input name="Submit" type="submit" value="提交" />
```

2. 评论功能的实现

博客的评论或者留言在博客系统中用得比较多。添加一条评论的代码如下：

```java
public class InsertComment{
  public static void insert(Comment comment)
  {
      DBConnectPool dbp=DBConnectPool.getInstance();
      Connection conn=dbp.getConnection();
      PreparedStatement pstmt=null;
      try {
        pstmt=conn.prepareStatement("inser into comment values(?,?,?,?)");
        pstmt.setInt(1, comment.getNewid());
        pstmt.setDate(2, (java.sql.Date)(new Date()));
        pstmt.setString(3,comment.getComPerson());
        pstmt.setString(4, content.getContext());
        pstmt.executeUpdate();
      } catch (SQLException e1) {
        e1.printStackTrace();
      }finally
      {
        try
        {
            if(pstmt!=null)
              pstmt.close();
        }catch(SQLException e1)
        {
            e1.printStackTrace();
        }
      }
  }
}</html>
```

3. 博客添加功能的实现

博客的管理包括博客的添加、删除和修改等操作。添加博客的相关代码如下：

```java
public static boolean insert(News news) {
    boolean flag = false;
    DBConnectPool dbp = DBConnectPool.getInstance();
    Connection conn = dbp.getConnection();
    PreparedStatement pstmt = null;
    try {
        pstmt = conn.prepareStatement(
            "insert into article(" +
            "Id,Title,classid," +
            "zhengwen,Blogid)" +
            " value(?,?,?,?,?);");
        pstmt.setString(2,news.getTitle());
        pstmt.setDate(3, news.getclassid());
```

```
                pstmt.setInt(4,news.getzhengwen());
                pstmt.setString(5, news.getBlogid());
                int r = pstmt.executeUpdate();
                if (r == 1)
                    flag = true;
            } catch (SQLException e1) {
                // TODO Auto-generated catch block
                e1.printStackTrace();
            } finally {
                try {
                    if (pstmt != null)
                        pstmt.close();
                } catch (SQLException e1) {
                    e1.printStackTrace();
                }
            }
            return flag;
        }
```

以上的 3 个功能的实现过程中会涉及大量的数据库操作，为了避免在页面中大量使用重复的代码，可以使用一个数据库的辅助类来实现数据库连接的操作。

## 11.2 网上书店

本实例用 Java Web 技术建立一个简单的电子商务应用系统——网上书店，我们采用 Sun 公司倡导的 JSP+JavaBeans 模式。

### 11.2.1 系统功能

•用户注册。新用户填写表单，填写用户名、E-mail 地址等信息。登录的用户名和密码不能为空。如果注册出错，提示出错信息，如果注册成功则进入图书购买页面。

•用户登录。输入用户名、密码。如果登录出错，系统将显示错误信息并且可以通过超级链接重新进入登录页面；如果登录成功，用户被链接到"订购图书"页面。

•书目浏览。成功登录的用户可以分页浏览图书书目，并将想要订购的图书提交到订购图书页面。如果用户直接进入该页面或没有成功登录就进入该页面，将被链接到"用户登录"页面。

•订购图书。成功登录的用户可以在该页面订购所需要的图书，可以选择自己需要的图书和图书的数量。如果用户直接进入该页面或没有成功登录就进入该页面，将被链接到"用户登录"页面。

•查看购物车。成功登录的用户可以在该页面查看自己的购物车，还可以在该页面修改或删除自己的购物车信息，可以修改图书的数量，也可以删除图书。如果用户直接进入该页面或没有成功登录就进入该页面，将被链接到"用户登录"页面。

•用户注销。成功登录的用户可以使用该模块退出登录。

### 11.2.2 数据库设计

我们采用纯 Java 驱动方式访问数据库。用 MySQL 建一个数据库 shop。该库中包括如下表。

（1）注册信息表 users。用户的注册信息存入数据库 shop 的 users 表中。users 表结构如表 11-5 所示。

表 11-5　users 表结构

| 字 段 名 称 | 数 据 类 型 | 说　　明 |
| --- | --- | --- |
| Username | varchar(20) | 用户登录名 |
| Password | varchar(20) | 口令 |
| name | varchar(30) | 口令 |
| sex | char(1) | 性别 |
| address | varchar(100) | 地址 |
| pcode | varchar(6) | 邮政编码 |
| email | varchar(30) | 电子邮件地址 |
| tel | varchar(20) | 联系电话 |
| regdate | datetime | 注册时间 |

（2）书目表 books。books 表用来存放图书书目。books 表结构如表 11-6 所示。

表 11-6　books 表结构

| 字 段 名 称 | 数 据 类 型 | 说　　明 |
| --- | --- | --- |
| Id | 自动编号 | 主键 |
| book_name | varchar(50) | 书名 |
| author | varchar(30) | 作者 |
| publish | varchar(50) | 出版社 |
| isbn | varchar(20) | ISBN 号 |
| content | varchar(4000) | 单价 |
| price | float(8,0) | 分类 |
| pubdate | datetime | 出版时间 |
| amount | int(4) | 订购数量 |
| leavecnt | int(4) | 剩余数量 |

### 11.2.3　系统的关键技术

**1. 生成数据库连接**

在该系统中无论是用户的注册登录，还是图书信息的查询，都需要用到数据库的连接操作。为了方便在各个模块中使用数据库连接，这里可以把数据库连接的代码单独拿出来作为一个类来完成，这样就可以重复使用它，提高系统的效率，也便于对代码进行修改。

```
DBConnect.java
package bean;
import java.sql.*;
public class DBConnect {
    Connection con=null;
    //JDBC 驱动程序名称
    String drivername="com.mysql.jdbc.Driver";
    //连接数据库 url
    String url="jdbc:mysql://localhost:3306/shop";
    //连接数据库用户名为
```

```java
    String username="root";
    //连接数据库密码为空
    String password="";
    public DBConnect() {
    }
public Connection getConnection(){
    try{
        //加载JDBC驱动程序
        Class.forName(drivername);
        //连接数据库
        con=DriverManager.getConnection(url,username,password);
        }catch(ClassNotFoundException e){
            e.printStackTrace();
        }catch(SQLException e){
            e.printStackTrace();
        }
    return con;
}
public static void main(String args[]){
    (new DBConnect()).getConnection();
}
}
```

2. 中文乱码的处理

在网站的开发过程中会涉及中文的显示，但是由于服务器还有浏览器编码等的不一致会造成中文乱码。在网上书店的开发过程中可以使用过滤器来集中实现乱码的处理，所有的请求在到达之前先进行字符集设置，通常是截获表单的提交请求。

注册页面使用的 bean 如下。

```java
EncodingFilter.java
package bean;
import java.io.*;
import javax.servlet.*;
import javax.servlet.http.*;
public class EncodingFilter implements Filter{
    String encoding = "GBK";
public void init(FilterConfig filterConfig)
        throws ServletException{
    encoding = filterConfig.getInitParameter("encoding");
}
public void doFilter(ServletRequest request,
        ServletResponse response,
        FilterChain chain)
        throws java.io.IOException,ServletException{
    request.setCharacterEncoding(encoding);
    chain.doFilter(request,response);
}
public void destroy(){}
}
```

配置信息如下：

```xml
<filter>
  <filter-name>Set Character Encoding</filter-name>
  <filter-class>bean.EncodingFilter</filter-class>
  <init-param>
      <param-name>encoding</param-name>
      <param-value>GBK</param-value>
  </init-param>
</filter>
<filter-mapping>
  <filter-name>Set Character Encoding</filter-name>
  <url-pattern>/*</url-pattern>
</filter-mapping>
```

该过滤器可以过滤所有的请求。

### 11.2.4 各个页面的设计

**1. 主页**

主页由导航条和一个欢迎语组成。

bookmain.jsp

```jsp
<%@ page contentType="text/html;charset=GB2312" %>
<HTML>
<BODY bgcolor =green>
<table align="center" border="0" width="740" height="18" bgcolor=
        yellow cellspacing="1">
<tr>
<td width="100%">
   <a  href="login.html">用户登录</a> |
   <a  href="reg.html"">用户注册</a> |
   <a  href="booklist.jsp">在线购书</a> |
   <a  href="cartlist.jsp">我的购物车</a> |
   <a  href="logout.jsp">用户注销</a> |
   </td>
   </tr>
</table>
   <H1>
    <CENTER> 欢迎光临网上书店</CENTER>
</BODY>
</HTML>
```

**2. 用户注册**

用户的注册信息需要存入数据库 shop 的 users 表中。

该模块需要使用的 bean 有两个，一个是 User.java，封装用户信息；另一个是 UserUtil.java，主要完成用户信息添加及查找用户功能。

## User.java

```java
package bean;
public class User {
    private String username;
    private String password;
    private String name;
    private char sex;
    private String address;
    private String pcode;
    private String email;
    private String tel;
//设置属性值、获取属性值的方法
    public String getUsername() {
        return username;
    }
    public void setUsername(String username) {
        this.username = username;
    }
    public String getPassword() {
        return password;
    }
    public void setPassword(String password) {
        this.password = password;
    }
    public String getName() {
        return name;
    }
    public void setName(String name) {
        this.name = name;
    }
    public char getSex() {
        return sex;
    }
    public void setSex(char sex) {
        this.sex = sex;
    }
    public String getAddress() {
        return address;
    }
    public void setAddress(String address) {
        this.address = address;
    }
    public String getPcode() {
        return pcode;
    }
    public void setPcode(String pcode) {
        this.pcode = pcode;
```

```java
        }
    public String getEmail() {
        return email;
    }
    public void setEmail(String email) {
        this.email = email;
    }
    public String getTel() {
        return tel;
    }
    public void setTel(String tel) {
        this.tel = tel;
    }
    public User() {
    }
    public User(String address, String email, String name, String password,
            String pcode, char sex, String tel, String username) {
        this.address = address;
        this.email = email;
        this.name = name;
        this.password = password;
        this.pcode = pcode;
        this.sex = sex;
        this.tel = tel;
        this.username = username;

    }
}
```
UserUtil.java:
```java
package bean;
import java.sql.*;
public class UserUtil {
        private Connection con;
    //查询数据库表中的用户信息是否存在
    public boolean findUser(String username, String password) {
        con=(new DBConnect()).getConnection();
        boolean flag =false;
        Statement stmt;
        ResultSet rs;

        String sql="select * from users where username='"+username+
                "' and password='"+password+"'";
        try{
            stmt=con.createStatement();
            rs=stmt.executeQuery(sql);
            if (rs.next())
                flag=true;
            rs.close();
```

```java
            stmt.close();
            con.close();
        }catch(Exception e) {
            e.printStackTrace();
        }
        return flag;
    }

    //向数据库表中添加一个user用户信息
    public boolean addUser(User user) {
        boolean flag=false;            //flag 标志
        if (user != null){
            con=(new DBConnect()).getConnection();
            Statement stmt;
            try{
                stmt = con.createStatement();
                String sql = "select * from users where username = '" +
                    user.getUsername() +"'";
                ResultSet rs = stmt.executeQuery(sql);
                if (rs.next()){
                    rs.close();
                    return flag;
                }
                sql = "insert into users(username,password,name,sex,address,tel,
                    pcode,email,regdate) values ('"
                        +user.getUsername()+"','"
                        +user.getPassword()+"','"
                        +user.getName()+"','"
                        +user.getSex()+"','"
                        +user.getAddress()+"','"
                        +user.getTel()+"','"
                        +user.getPcode()+"','"
                        +user.getEmail()+"',now())";
                stmt.execute(sql);
                flag=true;
                stmt.close();
                con.close();
            }catch(Exception e) {
                e.printStackTrace();
            }
        }
        return flag;
    }
}
```

该模块需要的页面有两个，一个是 reg.html，注册信息的提交；另一个是 reg.jsp，主要完成注册信息的处理。

注册信息的提交页面如下。

reg.html

```html
<html> <head><title>用户注册</title>
    <script>
        function openScript(url, name, width, height) {
            var Win = window.open(url, name, 'width=' + width + ',height='
                    + height + ',resizable=1, scrollbars=yes,
                    menubar=no, status=yes' );
        }
        function checkform() {
            if (document.form1.username.value==""){
                alert("用户名不能为空");
                document.form1.username.focus();
                return false;
            }
            if (document.form1.passwd.value==""){
                alert("用户密码不能为空");
                document.form1.passwd.focus();
                return false;
            }
            if (document.form1.passwd.value!=
                            document.form1.passconfirm.value){
                alert("确认密码不相符! ");
                document.form1.passconfirm.focus();
                return false;
            }
            return true;
        }
    </script>
  </head>
  <body>
    <div align="center">
  <p><h2>网上书店用户注册</h2></p>
  <form name="form1" method="post" action="reg.jsp" onSubmit=
            "return checkform();">
<table width="400" border="1" cellspacing="1" cellpadding=
            "1" bgcolor="#CCCCCC">
      <tr>
  <td colspan="2" align="center"><font face="黑体" size="4">用户注册
            </font></td>
      </tr>
      <tr>
        <td width="120" align="right">用户名：</td>
        <td width="280">
          <input type="text" name="username" maxlength="20" size="20" >
        </td>
      </tr>
      <tr>
```

```html
    <td width="120" align="right">密  码：</td>
    <td width="280">
      <input type="password" name="passwd" maxlength="20" size="20">
    </td>
</tr>
<tr>
    <td width="120" align="right">确认密码：</td>
    <td width="280">
      <input type="password" name="passconfirm" maxlength="20" size="20">
    </td>
</tr>
<tr>
    <td width="120" align="right">真实姓名：</td>
    <td width="280">
      <input type="text" name="name" maxlength="20" size="20">
    </td>
</tr>
<tr>
    <td width="120" align="right">性  别：</td>
    <td width="280">
      <input type="radio" checked="checked" name="sex" value="0">男
      <input type="radio" name="sex" value="1">女
    </td>
</tr>
<tr>
    <td width="120" align="right">电子邮件：</td>
    <td width="280">
      <input type="text" name="email" maxlength="50" size="25">
    </td>
</tr>
<tr>
    <td width="120" align="right">联系电话：</td>
    <td width="280">
      <input type="text" name="tel" maxlength="25" size="16">
    </td>
</tr>
<tr>
    <td width="120" align="right">联系地址：</td>
    <td width="280">
      <input type="text" name="address" maxlength="150" size="40">
    </td>
</tr>
<tr>
    <td width="120" align="right">邮政编码：</td>
    <td width="280">
      <input type="text" name="pcode" maxlength="6" size="6">
    </td>
</tr>
```

```html
      <tr>
        <td colspan="2" align="center">
          <input type="submit" name="Submit" value="注册">  
          <input type="reset" name="reset" value="取消">
        </td>
      </tr>
    </table>
  </form>
</div></body></html>
```

注册信息的处理页面如下。

reg.sp

```jsp
<%@ page language="java" import="java.util.*" pageEncoding="gb2312"%>
<html>
  <head>
    <title>用户注册</title>
  </head>
<%
    request.setCharacterEncoding("GB2312");
%>
<%@ page import="bean.User"%>
    <jsp:useBean id="userUtil" scope="page" class="bean.UserUtil"/>
    <%--<jsp:setProperty name="userUtil" property="*" />--%>
  <body>
<%
    String msg = "";            //msg 提示信息
    User user = new User();
    String username = request.getParameter("username");
    String password = request.getParameter("passwd");
    String name = request.getParameter("name");
    String sex = request.getParameter("sex");
    String address = request.getParameter("address");
    String pcode = request.getParameter("pcode");
    String tel = request.getParameter("tel");
    String email = request.getParameter("email");
    user.setUsername(username);
    user.setPassword(password);
    user.setName(name);
    user.setSex(sex.charAt(0));
    user.setAddress(address);
    user.setPcode(pcode);
    user.setTel(tel);
    user.setEmail(email);
    if(userUtil.addUser(user)) {
        session.setAttribute("username", user.getUsername());
        response.sendRedirect("booklist.jsp");
    }else{
        msg = "注册时出现错误,请稍后再试";
```

```
        }
%>
        <%= msg %>
    </body>
</html>
```

### 3. 用户登录

用户在该页输入自己的用户名和密码，系统对用户名和密码进行验证，如果用户身份合法将链接到订购图书页面，否则提示有误。本模块包含两个页面：登录页面和登录处理页面。登录页面如下。

login.html

```
<%@ page language="java" import="java.util.*" pageEncoding="gb2312" %>
<html>
  <head>
    <base href="<%= basePath %>">
    <title>用户登录检测</title>
  </head>
<jsp:useBean id="userutil" scope="page" class="bean.UserUtil" />
<%
        String msg = "";           //msg 信息
        String username =request.getParameter("username");
        String password = request.getParameter("password");
    if (userutil.findUser(username,password)){  //findUser 查询数据库表中
                                                  的用户信息是否存在
        session.setAttribute("username", username);//session 作用域,存储用户信息
        response.sendRedirect("booklist.jsp");
    }
    else{
        msg = "登录出错！";
    }
%>
  <body>
<div align="center">
    <%= msg %> <a href="Login.html">重新登录</a>
</div>
  </body>
</html>
```

登录处理页面如下。

login..jsp

```
<%@ page language="java" import="java.util.*" pageEncoding="gb2312" %>
<html>
  <head>
    <base href="<%= basePath %>">
    <title>用户登录检测</title>
  </head>
```

```jsp
<jsp:useBean id="userutil" scope="page" class="bean.UserUtil" />
<%
    String msg = "";              //msg 信息
    String username =request.getParameter("username");
    String password = request.getParameter("password");
 if (userutil.findUser(username,password)){  //findUser 查询数据库表中
                                              的用户信息是否存在
    session.setAttribute("username", username);  //session 作用域,存储用户信息
    response.sendRedirect("booklist.jsp");
    }
    else{
        msg = "登录出错! ";
    }
%>
  <body>
<div align="center">
    <%= msg %> <a href="Login.html">重新登录</a>
</div>
  </body>
</html>
```

4. 订购图书

成功登录的用户可以在该页面订购图书。booklist.jsp 页面显示所有的图书可以选择购买，选择购买后可以进入 purchase.jsp 进行处理，并且可以输入数量进行购买。

订购页面使用 Book.java 负责封装图书信息，使用 BookUtil.java 负责图书信息分页显示及查找图书信息功能，使用 Item.java 封装购物车的一个购买的条目信息，使用 Cart.java 实现购物车操作。

Book.java

```java
package bean;
import java.util.Date;
public class Book {
    private int id;
    private String bookname;
    private String author;
    private String publish;
    private String isbn;
    private String content;
    private float price;
    private Date pubdate;
    private int amount;
    private int leavecnt;
    public int getId() {
        return id;
    }
    public void setId(int id) {
        this.id = id;
```

```java
    }
    public String getBookname() {
        return bookname;
    }
    public void setBookname(String bookname) {
        this.bookname = bookname;
    }
    public String getAuthor() {
        return author;
    }
    public void setAuthor(String author) {
        this.author = author;
    }
    public String getPublish() {
        return publish;
    }
    public void setPublish(String publish) {
        this.publish = publish;
    }
    public String getIsbn() {
        return isbn;
    }
    public void setIsbn(String isbn) {
        this.isbn = isbn;
    }
    public String getContent() {
        return content;
    }
    public void setContent(String content) {
        this.content = content;
    }
    public float getPrice() {
        return price;
    }
    public void setPrice(float price) {
        this.price = price;
    }
    public Date getPubdate() {
        return pubdate;
    }
    public void setPubdate(Date pubdate) {
        this.pubdate = pubdate;
    }
    public int getAmount() {
        return amount;
    }
    public void setAmount(int amount) {
        this.amount = amount;
```

```java
    }
    public int getLeavecnt() {
        return leavecnt;
    }
    public void setLeavecnt(int leavecnt) {
        this.leavecnt = leavecnt;
    }
    public Book(int amount, String author, String bookname, String content,
            int id, String isbn, int leavecnt, float price, Date pubdate,
            String publish) {
        this.amount = amount;
        this.author = author;
        this.bookname = bookname;
        this.content = content;
        this.id = id;
        this.isbn = isbn;
        this.leavecnt = leavecnt;
        this.price = price;
        this.pubdate = pubdate;
        this.publish = publish;
    }
    public Book() {
    }
}
```

BookUtil.java 实现图书信息分页显示及查找图书信息功能:
BookUtil.java

```java
package bean;
import java.sql.*;
import java.util.Vector;
public class BookUtil {
        private Vector booklist;            //Vector 矢量，显示图书列表向量数组
        private int page = 1;          //显示的页码
        private int pageSize=10;           //每页显示的图书数
        private int pageCount =0;      //页面总数
        private long recordCount =0;       //查询的记录总数
        private Connection con;
        private Statement stmt;
        private ResultSet rs;
    public BookUtil() {
    }
    public Vector getBooklist() {
        return booklist;
    }
    public void setBooklist(Vector booklist) {
        this.booklist = booklist;
    }
    public int getPage() {
        return page;
```

```java
    }
    public void setPage(int page) {
        this.page = page;
    }
    public int getPageSize() {
        return pageSize;
    }
    public void setPageSize(int pageSize) {
        this.pageSize = pageSize;
    }
    public int getPageCount() {
        return pageCount;
    }
    public void setPageCount(int pageCount) {
        this.pageCount = pageCount;
    }
    public long getRecordCount() {
        return recordCount;
    }
    public void setRecordCount(long recordCount) {
        this.recordCount = recordCount;
    }
    //查询参数 Page 所指定的页的记录，并存放在向量数组 booklist 中
    public boolean execute(int page) {
        //取出记录数
        String sql = "select count(*) from books";
        try{
            con = (new DBConnect()).getConnection();
            stmt = con.createStatement();
            rs = stmt.executeQuery(sql);
            if (rs.next()) recordCount = rs.getInt(1);
            rs.close();
            stmt.close();
            con.close();
        }
        catch (Exception e){
            return false;
        }
        //设定有多少 pageCount
        if (recordCount < 1)
          pageCount = 0;
        else
          pageCount = (int)(recordCount - 1) / pageSize + 1;
        //检查查看的页面数是否在范围内
        if (page < 1)
          page = 1;
        else if (page > pageCount)
          page = pageCount;
```

```java
        //当前页显示 cnt 条记录
        int cnt;
        if (page==pageCount)
            cnt=(int)recordCount % pageSize;
        else
            cnt=pageSize;
        //当前页的第一条记录号
        int recordno=(page-1)*pageSize+1;
        //sql 为倒序取值
        sql = "select * from books order by id";
        try{
            con = (new DBConnect()).getConnection();
            stmt = con.createStatement(ResultSet.TYPE_SCROLL_INSENSITIVE,
            ResultSet.CONCUR_READ_ONLY);
            rs = stmt.executeQuery(sql);
            rs.absolute(recordno);
            booklist = new Vector();
            do
            {
                Book book = new Book();
                book.setId(rs.getInt("id"));
                book.setBookname(rs.getString("bookname"));
                book.setAuthor(rs.getString("author"));
                book.setPublish(rs.getString("publish"));
                book.setIsbn(rs.getString("isbn"));
                book.setContent(rs.getString("content"));
                book.setPrice(rs.getFloat("price"));
                book.setAmount(rs.getInt("amount"));
                book.setLeavecnt(rs.getInt("leavecnt"));
                book.setPubdate(rs.getDate("pubdate"));
                booklist.add(book);
                cnt--;
            }while (rs.next() && cnt!=0);
            rs.close();
            stmt.close();
            con.close();
            return true;
        }
        catch (SQLException e){
            return false;
        }
    }
    //查询参数 newid 所指定的图书,并返回查询结果
    public Vector getOnebook(int newid) {
        try
        {
            con = (new DBConnect()).getConnection();
            stmt = con.createStatement();
```

```java
            String sql="select * from books where id = " + newid ;
            rs = stmt.executeQuery(sql);
            if (rs.next())
            {
                booklist = new Vector();
                Book book = new Book();
                book.setId(rs.getInt("id"));
                book.setBookname(rs.getString("bookname"));
                book.setAuthor(rs.getString("author"));
                book.setPublish(rs.getString("publish"));
                book.setIsbn(rs.getString("isbn"));
                book.setContent(rs.getString("content"));
                book.setPrice(rs.getFloat("price"));
                book.setAmount(rs.getInt("amount"));
                book.setLeavecnt(rs.getInt("leavecnt"));
                book.setPubdate(rs.getDate("pubdate"));
                booklist.addElement(book);
            }
            rs.close();
            stmt.close();
            con.close();
            return booklist;
        }catch (Exception e){
            return null;
        }
    }
}
```

Item.java 封装购物车的一个购买的条目信息。

Item.java

```java
package bean;
public class Item {                    //Item 项目
        private long id;               //ID 序列号
        private long bookNo;           //图书表序列号
        private int amount;            //订货数量
    public long getId() {
        return id;
    }
    public void setId(long id) {
        this.id = id;
    }
    public long getBookNo() {
        return bookNo;
    }
    public void setBookNo(long bookNo) {
        this.bookNo = bookNo;
    }
    public int getAmount() {
```

```java
        return amount;
    }
    public void setAmount(int amount) {
        this.amount = amount;
    }
    public Item() {
        this.id=0;
        this.bookNo=0;
        this.amount=0;
    }
}
```

Cart.java 实现购物车操作：
Cart.java
```java
package bean;
import java.sql.*;
import java.util.Vector;
import javax.servlet.http.*;
public class Cart {
    private HttpServletRequest request;    //建立页面请求
    private HttpSession session;           //页面的 session
    private Vector purchaselist;           //显示图书列表向量数组
    private boolean isEmpty=false;         //库中的书数量是否够购买的数
    private int leaveBook=0;               //库存数量
    private Connection con;
    private Statement stmt;
    private ResultSet rs;

    public Vector getPurchaselist() {      //getPurchaselist()得到购买单
        return purchaselist;
    }

    public void setIsEmpty(boolean flag){
        isEmpty = flag;
    }
    public boolean getIsEmpty() {
        return isEmpty;
    }
    public void setLeaveBook(int bknum) {
        leaveBook = bknum;
    }
    public int getLeaveBook() {
        return leaveBook;
    }
    public boolean addnew(String id, String num, HttpServletRequest
            newrequest) throws Exception{
        request = newrequest;
        long bookid = 0;
```

```java
        int amount = 0;
try
{
    bookid = Long.parseLong(id);
    amount = Integer.parseInt(num);
}
catch (Exception e)
{
    return false;
}
if (amount<1) return false;
session = request.getSession(false);
if (session == null)
{
    return false;
}
purchaselist = (Vector)session.getAttribute("cart");
String sql = "select leavecnt from books where id=" + bookid;
try{
    con=(new DBConnect()).getConnection();
    stmt = con.createStatement();
    rs = stmt.executeQuery(sql);
    if (rs.next()){
        if (amount > rs.getInt(1)){
            leaveBook = rs.getInt(1);
            isEmpty = true;
            return false;
        }
    }
}
catch (SQLException e){
    return false;
}
Item iList = new Item();
iList.setBookNo(bookid);
iList.setAmount(amount);
boolean match = false;            //是否购买过该图书
if (purchaselist==null){          //第一次购买
    purchaselist = new Vector();
    purchaselist.addElement(iList);
}
else{ // 不是第一次购买
    for (int i=0; i< purchaselist.size(); i++) {
        Item itList= (Item) purchaselist.elementAt(i);
        if ( iList.getBookNo() == itList.getBookNo() ) {
            itList.setAmount(itList.getAmount() + iList.getAmount());
            purchaselist.setElementAt(itList,i);
            match = true;
```

```
                    break;
                }    //if name matches 结束
            }        // for 循环结束
            if (!match) {
                purchaselist.addElement(iList);
                
            }
        }
        session.setAttribute("cart", purchaselist);
        return true;
    }
    public boolean modiShoper(String id,String num,HttpServletRequest
            newrequest)throws Exception {
        request = newrequest;
        long bookid = 0;
        int amount = 0;
        try{
            bookid = Long.parseLong(id);
            amount = Integer.parseInt(num);
        }
        catch (Exception e){
            return false;
        }
        if (amount<1) return false;
        session = request.getSession(false);
        if (session == null){
            return false;
        }
        purchaselist = (Vector)session.getAttribute("cart");
        if (purchaselist==null){
            return false;
        }
        String sql = "select leavecnt from books where id=" + bookid;
        try
        {
            con=(new DBConnect()).getConnection();
            stmt = con.createStatement();
            rs = stmt.executeQuery(sql);
            if (rs.next()){
                if (amount > rs.getInt(1)){
                    leaveBook = rs.getInt(1);
                    isEmpty = true;
                    return false;
                }
            }
        }
        catch (SQLException e){
            return false;
```

```java
        }
        for (int i=0; i< purchaselist.size(); i++){
            Item itList= (Item) purchaselist.elementAt(i);
            if ( bookid == itList.getBookNo() ){
                itList.setAmount(amount);
                purchaselist.setElementAt(itList,i);
                break;
            }
        }
        return true;
    }

    public boolean delShoper(String delID,HttpServletRequest
                newrequest)throws Exception {
        request = newrequest;
        long bookid = 0;
        try{
            bookid = Long.parseLong(delID);
        }
        catch (Exception e){
            return false;
        }
        session = request.getSession(false);
        if (session == null){
            return false;
        }
        purchaselist = (Vector)session.getAttribute("cart");
        if (purchaselist==null){
            return false;
        }
        for (int i=0; i< purchaselist.size(); i++) {
            Item itList= (Item) purchaselist.elementAt(i);
            if ( bookid == itList.getBookNo() ) {
                purchaselist.removeElementAt(i);
                break;
            }
        }
        return true;
    }
}
```

图书显示页面如下。

**booklist.jsp**

```jsp
<%@ page language="java" import="java.util.*" pageEncoding="gb2312"%>
<%@ page import="bean.Book"%>
<jsp:useBean id="bookutil" scope="page" class="bean.BookUtil" />
<%
        response.setCharacterEncoding("gb2312");
```

```jsp
    if(session.getAttribute("username")==null){
        response.sendRedirect("Login.jsp");
    }
%>
<html>
  <head>
    <title>网上书店--选购图书</title>
  </head>
<%
    String requestpage ="";
    if (request.getParameter("page")==null || request.getParameter("page").equals("")) {
        requestpage="1";
    }else{
        requestpage = request.getParameter("page");
    }
    int pageno = Integer.parseInt(requestpage);//pageno 当前页码
    bookutil.setPage(pageno);
%>
  <body>
    <div align="center">
    <table width="800" border="0" cellspacing="1" cellpadding="1">
    <tr>
        <td width="100"> </td>
        <td width="150"><a href="booklist.jsp">在线购物</a></td>
        <td width="150"><a href="cartlist.jsp">我的购物车</a></td>
        <td width="150">当前用户：<%=
            (String)session.getAttribute("username") %></td>
        <td width="250"><a href="logout.jsp">用户注销</a></td>
    </tr>
    <tr>
        <td colspan="3"></td>         <!--colspan 和并列-->
    </tr>
    </table>
    <table width="600" border="0" cellspacing="1" cellpadding="1">
    <tr valign="center">
    <td height="40" align="center"><span class="booktitle">
        网上书店图书列表</span></td>
    </tr>
    <tr>
        <td>
        <table width="100%" border="0" cellspacing="0" cellpadding="0" >
<%
    if (bookutil.execute(pageno)) {
        if (bookutil.getBooklist().size() > 0 ){
            for (int i=0; i < bookutil.getBooklist().size(); i++){
                Book book = (Book)bookutil.getBooklist().elementAt(i);
%>
        <tr>
```

```
                    <td width="100" height="30">图书名称:</td>
                    <td><span class="bookname_style"><%= book.getBookname()
                        %></span></td>
                </tr>
                <tr>
                    <td height="20">ISBN:</td>
                    <td height="20"><%= book.getIsbn() %></td>
                </tr>
                <tr>
                    <td height="20">作者:</td>
                    <td height="20"><%= book.getAuthor() %></td>
                </tr>
                <tr>
                    <td height="20">出版社:</td>
                    <td height="20"><%= book.getPublish() %></td>
                </tr>
                <tr>
                    <td height="20">单价:</td>
                    <td height="20">¥<%= book.getPrice()%>元</td>
                </tr>
                <tr>
                    <td height="20">出版时间:</td>
                    <td height="20"><%= book.getPubdate()%></td>
                </tr>
                <tr>
                    <td colspan="2" align="right"><a href="purchase.jsp?bookid=
                        <%= book.getId() %>" >我要购买</a></td>
                </tr>
                <tr>
                    <td colspan="2"><hr></td>
                </tr>
<%          }
        }else {
            out.println("<tr><td align='center' colspan=6> 
                暂时没有此类图书资料</td></tr>");
        }
} else {
%>
                <tr>
                    <td align="center" colspan=6> 数据库出错,请稍后</td>
                </tr>
<% } %>
        </table>
每页10条信息,共<%=bookutil.getRecordCount()%>条 第<%=bookutil.getPage()
            %>页 共<%=bookutil.getPageCount()%>页
<br>
    <table>
    <tr>
```

```jsp
            <td>
<%
    if(bookutil.getPage()==1){
        out.print(" 首页 上一页");
    }else{
%>
        <a href="booklist.jsp?page=1">首页</A>
        <a href="booklist.jsp?page=<%= bookutil.getPage()-1 %>">上一页</A>
<%
    }
%>
<%
    if(bookutil.getPage()>=bookutil.getPageCount()){
        out.print("下一页 尾页");
    }else{
%>
        <a href="booklist.jsp?page=<%=bookutil.getPage()+1%>">下一页</A>
        <a href="booklist.jsp?page=<%= bookutil.getPageCount()%>">尾页</A>
<%
    }
%>
转到第<select name="pageno" onChange="javascipt:gopage()">
<%
    for(int i=1; i<=bookutil.getPageCount(); i++)  {
    if (i== bookutil.getPage()){
%>
    <option selected value=<%=i%>><%=i%></option>
<%
        }else{
%>
        <option value=<%=i%>><%=i%></option>
<%
        }
    }
%>
    </select>页
        </td>
    </tr>
    </table>
      </td></tr>
    </table>
    </div></body></html>
```

purchase.jsp 订购数量输入页面如下。
Purchase.jsp
```jsp
<%@ page language="java" import="java.util.*" pageEncoding="gb2312"%>
<%@ page import="bean.Book" %>
<jsp:useBean id="bookutil" scope="page" class="bean.BookUtil" />
<jsp:useBean id="cart" scope="page" class="bean.Cart" />
```

```jsp
<%
    if(session.getAttribute("username")==null){
        response.sendRedirect("Login.html");
    }
%>
<html>
  <head>
    <title>网上书店--购买图书</title>
    <script language="javascript">
        function openScript(url,name, width, height)
        {
            var Win = window.open(url,name,'width=' + width + ',height=' + height + ',
                resizable=1,scrollbars=yes,menubar=no,status=yes');
        }
        function check()
        {
            if (document.form1.amount.value<1){
                alert("你的购买数量有问题");
                document.form1.amount.focus();
                return false;
            }
            return true;
        }
    </script>
  </head>
<%
    String msg = "";
    String submits = request.getParameter("Submit");
    int id=0;
    if (submits!=null && !submits.equals("")){
        String bookid = request.getParameter("bookid");
        String amount = request.getParameter("amount");
        if (cart.addnew(bookid,amount,request)){
            msg = "你需要的图书已经放入你的购物车中！";
        }else if (cart.getIsEmpty()){
            msg = "库存图书数量不足！只剩"+cart.getLeaveBook()+"本";
        }else {
            msg = "暂时不能购买！";
        }
    }else {
        if (request.getParameter("bookid")==null ||
                request.getParameter("bookid").equals("")) {
            msg = "你购买的图书不存在！";
        } else {
            try {
                id = Integer.parseInt(request.getParameter("bookid"));
                if (bookutil.getOnebook(id)==null){
                    msg = "你要购买的图书不存在！";
```

```
                    }
                } catch (Exception e){
                    msg = "你要购买的图书不存在！";
                }
            }
        }
%>
  <body onLoad="javascript:window.focus();">
<div align="center">
  <p><br></p><p>网上书店欢迎你选购图书！</p>
   <% if(!msg.equals("")){
        out.println(msg);
      } else {
        Book bk = (Book) bookutil.getBooklist().elementAt(0);
    %>
<table width="90%" border="0" cellspacing="2" cellpadding="1">
   <form name="form1" method="post" action="purchase.jsp">
     <tr>
       <td align="center">图书名：<%= bk.getBookname() %></td>
     </tr>
     <tr align="center">
       <td>你想要的数量：
         <input type="text" name="amount" maxlength="4" size="3" value="1">本</td>
     </tr>
     <tr align="center">
       <td>
        <input type="hidden" name="bookid" value="<%=request.getParameter('bookid')%>">
        <input type="submit" name="Submit" value="购 买"
                  onClick="return(check());">
         <input type="reset" name="Reset" value="取 消">
       </td>
     </tr>
   </form>
  </table>
<% } %>
   <br>
   <p><a href="javascript:window.close()">关闭窗口</a></p>
 </div>
   </body>
</html>
```

5. 查看修改购物车信息

查看修改购物车信息页面如下。

cartlist.jsp

```
<%@ page language="java" import="java.util.*" pageEncoding="gb2312"%>
<%@ page import="bean.*" %>
<jsp:useBean id="bookutil" scope="page" class="bean.BookUtil" />
```

```jsp
<jsp:useBean id="cart" scope="page" class="bean.Cart" />
<%
    if(session.getAttribute("username")==null){
        response.sendRedirect("Login.html");
    }
%>

<html>
  <head>
    <title>网上书店--我的购物车</title>
<script language="javascript">
function openScript(url,name, width, height){
    var Win = window.open(url,name,'width=' + width + ',height=' + height
            + ',resizable=1,scrollbars=yes,menubar=no,status=yes');
}
function checklogin() {
    if (document.payout.userid.value=="")
    {
        alert("你还没有登录,请登录后再提交购物清单。");
        return false;
    }
    return true;
}
function check()
{
    if (document.change.amount.value<1){
        alert("你的购买数量有问题");
        document.change.amount.focus();
        return false;
    }
    return true;
}
</script>
  </head>
<%
    String userid = (String) session.getAttribute("userid");
    if ( userid == null )
        userid = "";
    String modi = request.getParameter("modi");
    String del = request.getParameter("del");
    String clearCar = request.getParameter("clear");
    String msg = "";
    if (modi!=null && !modi.equals("")) {
        String id = request.getParameter("bookid");
        String amount = request.getParameter("amount");
        if (!cart.modiShoper(id,amount,request) ){
            if (cart.getIsEmpty())
                msg = "你要的修改购买的图书数量不足你的购买数量!";
```

```jsp
            else
                msg = "修改购买数量出错！";
        } else {
            msg = "修改成功";
        }
    }else if ( del != null && !del.equals("") ) {
        String delID = request.getParameter("bookid");
        if ( !cart.delShoper(delID,request) ) {
            msg = "删除清单中的图书时出错！" ;
        }
    }
%>
  <body>
<div align="center">
    <table width="750" border="0" cellspacing="1" cellpadding="1">
    <tr>
    <td width="200"> </td>
        <td width="100"><a href="booklist.jsp">在线购物</a></td>
        <td width="100"><a href="cartlist.jsp">我的购物车</a></td>
        <td><a href="logout.jsp">用户注销</a></td>
    </tr>
    </table>

    <table width="900" border="0" cellspacing="1" cellpadding="1">
    <tr valign="top">
    <td align="center">
        <p><br>
        <b><font color="#0000FF"><%=(String)session.getAttribute
            ("username")%>的购物车物品清单</font></b></p>
<%
    if (!msg.equals("") )
        out.println("<p ><font color=#ff0000>" + msg + "</font></p>");

        Vector cartlist = (Vector) session.getAttribute("cart");
        if (cartlist==null || cartlist.size()<0 ){
            if (msg.equals(""))
            out.println("<p><font color=#ff0000>你还没有选择购买图书！
                请先购买</font></p>");
        } else {
%>
      <table width="100%" border="1" cellspacing="1" cellpadding="1" >
        <tr align="center">
          <th width="180">图书名称</td>
          <th width="100">作者</th>
          <th width="160">书号</th>
          <th width="130">出版社</th>
          <th width="80">单价(元)</th>
          <th width="90">出版时间</th>
```

```jsp
            <th width="60">数量</th>
            <th colspan =2  width="100">选择</th>
        </tr>
    <%
    float totalprice =0;
    int totalamount = 0;
    for (int i = 0; i < cartlist.size(); i++){
        Item iList = (Item) cartlist.elementAt(i);
         if (bookutil.getOnebook((int)iList.getBookNo())!=null) {
             Vector new_book_list = bookutil.getOnebook((int)iList.getBookNo());
             Book book = (Book) new_book_list.elementAt(0);
             totalprice = totalprice + book.getPrice() * iList.getAmount();
             totalamount = totalamount + iList.getAmount();
    %>
        <tr>
            <td align="center"><%= book.getBookname() %></td>
           <td align="center"><%= book.getAuthor() %></td>
             <td align="center"><%= book.getIsbn() %></td>
            <td align="center"><%= book.getPublish() %></td>
            <td align="center"><%= book.getPrice() %></td>
            <td align="center"><%= book.getPubdate() %></td>
             <form name="change" method="post" action="booklist.jsp"
                    onSubmit="return check();">
            <td align="center">
              <input type="text" name="amount" maxlength="4" size="3"
                      value="<%= iList.getAmount() %>" >
            </td>
            <td align="center" width="55" ><form name="del" method="post"
                   action="cartlist.jsp" >
      <input type="hidden" name="bookid" value="<%= iList.getBookNo() %>" >
                <input type="submit" name="modi" value="修改" >
                </td>
             </form>
            <form name="del" method="post" action="cartlist.jsp" >
            <input type="hidden" name="bookid" value="<%= iList.getBookNo() %>" >
<td align="center" width="55"><input type="submit" name="del" value=
   "删除"></td>
             </form>
        </tr>
         <% }
   } %>  <tr><td colspan="9" align="right"><br>你选择的图书的总金额:<%=
        totalprice%>元  总数量: <%= totalamount%>本 </td></tr>
    </table>
      <p></p>
        <table width="90%" border="0" cellspacing="1" cellpadding="1">
          <tr>
            <td align="right" valign="bottom"> <a href="booklist.jsp">
              继续购书</a>   
```

```
                </td>
            </tr>
        </table>
<% } %>
        </td>
    </tr>
</table>   </div></body></html>
```

6. 用户注销

该模块负责销毁用户的 session 对象，导致登录失败。

logout.jsp:

```
<%
    session.invalidate();         //invalidate 使无效
    response.sendRedirect("Login.html");
%>
```

## 11.3 小　　结

本章具体介绍了两个应用系统的编程思路以及实现方法。它们综合运用了 Java Web 的技术和方法。通过对本章的学习，加深了对 Java Web 相关概念、方法和技术的理解和灵活运用。

## 11.4 习　　题

1. 请用 Java Web 技术开发一个校园网通讯录系统。
2. 以你所在的学校的图书管理业务为例，请用 Java Web 技术开发网上图书管理系统。
3. 完成本章两个实例的上机编码、调试、运行测试实训。

# 参 考 文 献

Liang Y D. 2004. Java 语言程序设计（基础篇）. 北京：机械工业出版社
Liang Y D. 2006. Java 语言程序设计（进阶篇）. 北京：机械工业出版社
Wang P S. 2003. Java 面向对象程序设计. 北京：清华大学出版社
Arnold K, Gosling J, Holmes D. 2005. The Java Programming Language. 4th ed. Upper Saddle River: Addison-Wesley
Eckel B. 2006. Thinking in Java. 4th ed. Boston: Prentice Hall